KINE1000 Sociocultural Perspe

Critical Skills Manual
For undergraduate students of kinesiology

Nick Ashby

Copyright © 2008 by McGraw-Hill Ryerson Limited

No part of this publication may be reproduced, stored in a retrieval system, or transmitted, in any form or by any means, electronic, mechanical, photocopying, recording, or otherwise without prior written permission from McGraw-Hill Ryerson Limited.

ISBN-10: 0-07-012321-7
ISBN-13: 978-0-07-012321-2

Printed and bound in Canada

Contents

1. **INTRODUCTION: CRITICAL SKILLS AND YOUR SUCCESS**   1
   1.1 University compared to high school   1
   1.2 Introducing success-generating skills   2-3

2. **EXAM SKILLS AND STRATEGIES**   4
   2.1 Exam preparation skills   4
   2.2 Strategies during the exam revision period   5-7
   2.3 Strategies when in the exam   8-10
   2.4 Other exam issues   10

3. **TIME-MANAGEMENT SKILLS**   11
   3.1 Creating a term schedule   11-12
   3.2 Creating a monthly schedule   13
   3.3 Creating a weekly schedule   14
   3.4 Creating task assessments for assignments   15-17

4. **ACTIVE LISTENING SKILLS**   18
   4.1 Active listening versus passive hearing   18
   4.2 How to become an active listener   19-21

5. **NOTE-TAKING SKILLS**   22
   5.1 Mind map lecture notes   22-25
   5.2 Cornell style lecture notes   26-28

6. **ACTIVE READING SKILLS**   29
   6.1 Different kinds of course readings   29-30
   6.2 Active reading techniques   30-31
   6.3 Active reading exercise   32-39

7. **RESEARCH SKILLS**   40
   7.1 Read and understand assignment instructions   40-43
   7.2 Define the topic   43-45
   7.3 Do background research   45-52
   7.4 Locate suitable sources   53-56
   7.5 Select suitable sources   56-57
   7.6 Evaluate your selected sources   58-59

8. THE WRITING PROCESS	60
8.1 First draft	60-62
8.2 Second draft	62-63
8.3 Final draft	64-73
9. ESSAYS	74
9.1 The introduction	74-76
9.2 The body	76-80
9.3 The conclusion	81
9.4 Theses and thesis-statements	82-85
10. SOME OTHER KINDS OF ACADEMIC WRITING	86
10.1 Writing an abstract	86-87
10.2 Writing a synopsis	88
10.3 Writing an annotated bibliography	89
10.4 Writing a case study	90
10.5 Writing a scientific report	91
11. APA CITATIONS AND REFERENCES	92
11.1 In-text citations	92-100
11.2 Reference lists	100-110
11.3 Resources	110
12. CLASS PRESENTATION SKILLS	111
12.1 Researching your topic and writing the script	111-112
12.2 Preparing to give your presentation	113
12.3 On the day of your presentation	114-115
12.4 Using PowerPoint in a presentation	116-117

1. INTRODUCTION: CRITICAL SKILLS AND YOUR SUCCESS

1.1 University compared to High School

It has sometimes been said that university students are a different species of human beings. Like everyone else, they have too much to do and too little time in which to do it. Unlike others, however, university students are under no external pressure to complete their tasks. There are no tyrannical bosses or omnipresent high school teachers breathing down one's neck to ensure that things get done right, or done at all!

High school teachers not only grade assignments but they monitor progress and provide ongoing remedial classes and instruction for struggling students. **University professors are not quite the same as high school teachers.** The job of the university professor is not only to teach but also to engage in research and publish one's work. This means that your professors may not have time to monitor your progress (other than keep a record of your grades) or to actively intervene to provide remedial instruction if you struggle with your assignments because of undeveloped study skills. The assumption is that if you are struggling you will be proactive and work that much harder to succeed because you are an adult and fully responsible for your own decisions and destiny. University offers you the freedom to fail as well as the freedom to succeed. This is not to say that you won't encounter wonderful professors who will go out of their way to spend time helping you. But it is important to realize that the level of monitoring and support that you have been used to at high school is not generally available. In summary then, as a first year student just starting out, it is easy to be blown away by a newfound sense of freedom (lack of continual monitoring and nobody to pressure you into doing anything), yet hard to adjust to the lack ongoing support for your academic work if you are struggling. As you will no doubt realize, there is the potential for a negative first-year experience if these realities are not understood and planned for from the start.

One strategy for accomplishing the adjustment to university life and finding a comfort zone as an efficient, successful student is to develop your critical skills. Critical skills enable you to function autonomously at a high level and to do well on assignments and exams. Critical skills enable you to organize and carry out all your activities as a university student, both on and off campus; you might say that critical skills provide the framework for a successful 'academic lifestyle'. The purpose of this manual is to present you with the basics of some important critical skills that you will need in order to develop an academic lifestyle and to be highly successful in your academic endeavours.

1.2 Introducing Success-Generating Skills

Time is your most precious resource. When someone donates blood, the body replenishes it. But used up time from your life is never replaced – it is gone forever. So time is more precious even than life-blood! **Time management skills** are essential to being able to use your precious time well. It is crucial that you learn to use your time effectively in order to negotiate and successfully manage the many, often maddening simultaneous demands that will be placed on you. A common grumble of students is that they have lots of essays to complete by the same deadline, or several exams to revise for at once. You will sometimes hear a student say in an angry tone something like: "Don't my professors (or the university) realize how unreasonable it is to expect me to complete all these essays, or revise for all these exams at the same time. The scheduling is all screwed up!" But the reality is that the scheduling is not all screwed up. There is a lot of pressure on students to complete multiple assignments or revise at the same time for different courses. This is a part of reality at university. What makes this situation seem unreasonable as opposed to just very challenging is lack of time management skills. Procrastination, failing to regard time as well spent if not spent on entertainment and socializing, leaving things to the last minute, failing to identify goals and define properly scheduled steps for achieving them all contribute to a sense of being overwhelmed by unrealistic demands. Developing your time management skills will help you to avoid experiencing this negative illusion and to cope well with the pressures of student life.

If you ask people what the most common academic student activity is, they almost invariably say that it is attending lectures, writing essays or reading books. But this is not quite right. Talking and listening are the most common activities. Students talk about lectures, readings, and assignments (and about their professors, TAs, and each other!) both formally and informally inside and outside the classroom setting. Campus is a noisy place where the exchange of ideas is going on all the time. But talking and listening are like visual and auditory perception. Vision and hearing seem easy on the surface (just open your eyes and ears!), yet they are actually complex processes. Similarly, presenting thoughts and ideas vocally seems like an effortless activity – just open your mouth and speak. But the effective vocal presentation of ideas and viewpoints actually depends on important principles of clarity and organization of information. These principles are also relevant to the visual presentation of ideas. **Presentation skills** enable you to communicate your thoughts and ideas effectively. This will be important when you are asked to give a vocal or PowerPoint presentation in tutorial.

Listening also seems like an effortless activity – just be sure to catch what is said! But if you tried to notice and record literally everything said at lecture, you would soon be burnt out as there would be too much information to catch. You would also not learn anything from lecture because you would be too busy trying to write everything down! To be an effective listener at lectures and tutorials, you need to listen for important ideas and points while disregarding irrelevant information. Indeed, the same applies to intelligent conversation with friends and colleagues. Listening effectively requires **active listening skills**.

An important skill closely associated with active listening is note-taking. Lecture notes should not consist of a bunch of hastily written (or typed) scribble capturing an indiscriminate mish-mash of things said by the lecturer. Your lecture notes don't have to be neat, as they can always be tidied up later. But they should be meaningful to you concerning the lecture topic even as they are first written, and later they should function as a tool for further development of your understanding of the lecture topic. This is what **note-taking skills** are all about.

A lot of your time will be spent reading and trying to understand assigned course readings. As with listening, there is an effective and an ineffective way of reading. Ineffective reading proceeds according to the assumption on the part of the reader that everything in a book or article is of equal importance – hence, the purpose of reading is to remember the information expressed by every single sentence! This way of reading is ineffective because it is both time-consuming and does not actually promote memory or understanding. Its ineffectiveness can lead to frustration and despair, leaving students with the feeling that they are not smart enough, or that they are being buried under an insurmountable mountain of assigned readings. It is surprising how many students come up to university both with this useless approach to reading and with an attitude of reluctance to give it up! But give it up you must, if you are to become an effective reader and get the most out of course material. **Active reading skills** are all about giving up ineffective reading and about learning strategies for engaging with written material in a way that promotes genuine understanding, something that is highly prized – and highly rewarded!

It goes without saying that you will do a lot of written assignments during the course of your degree program. Students may struggle with written work for a number of reasons. They may have difficulty with grammar or vocabulary. Professors assume that students can read and write well simply by virtue of having made it to university. But the reality is that for a variety of factors students may have some difficulties. Professors are unlikely provide remedial help with this, but students may be able to access some limited support through a campus writing centre. There is another way in which one may struggle with writing, and this has more to do with the writing skills focused on in this manual. Students may have only a vague or even wrong understanding of the organization and function of different types of written work. For example, an essay has a different organization and function from a case study, reflection piece, or research report. **Writing skills** are all about sharpening your understanding of different types of writing and empowering you to put this into practice to produce effective written work.

For some types of written assignment, research beyond the assigned readings will be necessary. Researching sources involves developing an understanding of the ways of thinking about a topic or issue by experts or concerned parties. The quality of your research effort will be reflected in part by the quality, breadth and depth of the sources you cite and reference, as well as by how well you cite and reference them. **Research skills** (including citing and referencing sources) will help you to conduct and present your research in a timely and effective way.

Please bear in mind that nothing worthwhile is ever accomplished easily. This is certainly true of critical skills. They require determined and consistent practice. If you learn and practice these skills now, just as you are starting out, you will exceed your expectations and enjoy a very positive and successful academic experience. You will also be fully prepared for exams, which brings us to our next topic.

2. EXAM SKILLS AND STRATEGIES

Notice that the title of this section is not "Exam Revision." To revise means to read or look over something again. But depending on what the exam is testing you for, revision in this sense may not be the best strategy. Indeed, the concept of 'exam revision' is rather misleading because it suggests that there is some *special way* in which the student engages with course material around exam times but not at any other time in the academic year. This way of thinking may not lead to effective exam preparation, and it can be a recipe for disaster. However, there are some strategies that can be employed at exam time. Both exam skills and exam strategies are discussed below.

2.1 Exam Preparation Skills

The reality is that **everything you do for a course during the term is preparation for its exam**. That is why this chapter on exam skills is here toward the beginning of this manual. It is here at the beginning to impress upon you the urgent message that in order to be in good shape for an exam, you need to develop and use the skills explained in the subsequent pages of the manual. In particular, you need to employ time management skills to ensure that you arrange for regular, adequate study time every week to read assigned course materials and complete your lecture notes. You need to employ active listening skills in order to grasp the meaning and wider significance of what is said at lecture. Related to this, lecture note-taking skills are crucial to being able to take effective, meaningful notes that reflect active listening and help to develop your understanding. Active reading skills are essential to grasping the meaning and wider significance of course readings. **These are the essential exam preparation skills, and they are employed at the beginning of your courses and used all the way through the term**. It is not necessary to discuss these skills further here, as they are explained in detail in subsequent sections of this manual. Summing up:

Exam Preparation Skills

- **TIME MANAGEMENT for regular, adequate study periods every week**

- **ACTIVE LISTENING at lectures and tutorials in order to understand the meaning and wider significance of what is said or presented**

- **EFFECTIVE NOTE-TAKING at lecture and tutorial in order to create meaningful notes that help to develop your understanding**

- **ACTIVE READING in order to understand the meaning and wider significance of course readings.**

Start employing these exam preparation skills today, if not before!

2.2 Strategies during the Exam Revision Period

The following strategies are for those who, throughout the term, have employed the exam preparation skills indicated in section 2.1.

Get Clear On the Materials You Need to Revise

Before you begin to revise, you need to be clear about which materials are relevant. Obviously, assigned course readings will be, but is there anything else than you need to look at? Are there any lecture PowerPoint notes that you are expected to revise? Have there been any guest speakers, and will you be tested on their lectures in the exam?

Revising Course Readings and Lecture Notes

Revising course readings and lecture notes is, perhaps, the oldest and most common strategy used by students to work them up into a sense of readiness for an exam. It is also one of the most misunderstood strategies. Careful, detailed study of *all* of your course readings is not feasible a week before an exam – it is just not possible to compress a whole term or year's worth of study into a week or two – especially so when taking into account that this may be just one exam among several around the same time for different courses! But a close reading of all of your course materials should not be necessary, in any case. You have carefully read and studied all the course readings throughout the term. So, how should you revise, and why should you revise?

Make a Revision Schedule

Using time management skills (see section 3 of this manual), make a schedule for revision periods each day throughout the weeks before the exam is scheduled. This will probably only require you to convert timeslots in your term calendar to revision periods, as most courses and lecture attendance end before the official exam revision period begins. Out of the available time, schedule more free timeslots for selective revision than for scan revision (see below).

Selective Revision

It is seldom the case that all topics, issues, concepts or techniques covered by a course will be equally clear. Similarly, some course readings will be easier to understand than others. Selective revision is a useful strategy that you can use to focus your time and resources on what you regard as weak spots in your understanding that remain after the course has finished but before scheduled exams. Think about the course overall and make a list with two columns. In the first column, list the course topics that you feel reasonably comfortable about. In the second column, list the particular topics that you feel you have still not grasped. Revise the readings and lecture notes for the topics in the second column by carefully reading the material using active reading skills (see section 6 of this manual), or by practicing the techniques you're not clear about, if, for example, you are revising for a math course. Any remaining time before the exam can be used for scan revision (see below).

Scan Revision

Scan revision involves quickly reading over course readings and lecture notes to refresh one's memory of facts and details. Scan for main ideas, main issues, main connections between ideas, main principles, or main techniques. After completing a scan of each reading, close the book or turn away and write a list of keywords or key phrases from memory. Keep it brief. Then check your list against the reading and add keywords for any missed main points. Give the list a title connecting it to the course topic you have just revised. Bearing in mind the primary importance of selective revision, try to arrange your revision schedule so that you can scan revise systematically from the earliest to the latest course topics and materials. If there is any time left, use your lists of keywords for further revision. Work systematically from keyword lists for the earliest topics to those for the last topics of the course. Begin by looking at keywords on your list and recall everything you can in connection with them. Then scan the readings and lecture notes to see if there is anything important that you omitted. Add new, and this time circled keywords to your list for anything important that was missed.

<u>Scan Revision as an Anxiety Producer</u>

The human brain copes with a world of complex stimuli by filtering, categorizing and memorizing information unconsciously. Most of our memories are unconscious, and items of memory tend only to become conscious when triggered by external stimuli. For example, until mentioned in this sentence right now, your memory of your home address and postcode has likely been entirely unconscious for quite a while. But the fact that you did not consciously remember your home address until mentioned just now does not mean that you don't know where you live! Similarly, the fact that you do not consciously remember details about a course topic until the moment of scan revising the readings and lecture notes does not mean that you do not know about the topic. Some students who don't understand how memory works get spooked out by scan revising because it can highlight the fact that the information was not in one's consciousness. They mistakenly think that you only know something if you are conscious of it all the time. For the most part, you do not learn much about a topic from scan revision. Rather, scan revising is a way of confirming your knowledge by making you conscious of what you know (this is called "refreshing" one's memory). Scan revising can be a good was of generating self-assurance about your readiness for the exam. In the exam itself, the exam questions act as triggers for your memory in place of course readings and lecture notes.

Forming a Study Group

Schedule regular meetings with friends or colleagues interested in being part of a study group. Get together to plan an agenda, so that each meeting focuses on an agreed-upon range of course topics. Study groups can be arranged along selective or scan revision lines. For selective revision, members of the group should agree to carefully read the material in advance of the meeting, highlighting problematic places in the text and writing out any questions. Members would then attend the meeting, bringing their highlighted readings and questions with them to discuss with the rest of the group. Or, for longer group study sessions, members could agree to get together to carefully read the text and raise questions for the group as they arise during reading. For scan revision, members could each bring their reading material and scan it in each others company. Each member could test the others by turning her or his list of keywords and phrases into questions for the group. Being part of a study group is an excellent way of revising for an exam because no one person ever thought of every angle on a topic. Studying with others allows you to share their perspectives and enrich your own understanding.

Dealing with Anxiety

Even if you have employed all the exam preparation skills and are 100% ready, impending exams can create anxiety. It is only natural to be more stressed than usual around exam times. Anxiety is usually generated by fantasies of worst-case scenarios and worst-case reactions of others to failure. If, throughout the term you have employed the exam preparation skills outlined in section 2.1, *you will not fail the exam*. It is as simple as that! However, if you are an inveterate worrier, some of the following strategies may be helpful.

Schedule a Daily Worry Period

Arrange a one-hour worry period every day in your revision schedule. Get into the mindset of worrying about exams only during these periods. Make your worry periods worthwhile. Try to fantasize about very worst things that could happen if you failed the exam, allow yourself to pace up and down, be restless, cry if you want to, and sweat with the worry of it all. Crystallize each day's worth of worry into this period. Using worry periods can help you train yourself to deflect worry from times of the day when you need to concentrate on revision instead of on stressing yourself out.

Put the Exam into Perspective

There are very few teachers, professors, or other highly successful people who have not failed or done poorly in an exam at one time or another. Life happens, and we are all human. You will not fail your exam if you follow the advice in this manual, *but even if you did fail the exam*, your fantasies of worst-case reactions from others would still only be fantasies. Parents and friends would still love you. The world would not come to a crashing halt. And you would still have a future. How much of your total degree is the exam worth, anyhow – 1%, less?! Putting your exam into perspective won't stop you worrying, but it may stop your worry from getting in the way as you revise.

Clinical Anxiety

For some students, the exam period can trigger anxiety disorders, such as panic attacks. A panic attack is not simply a moment of intense worry. It is a moment of terror, with an overwhelming feeling that you are out of control and that something terrible is going to happen to you. Typically, this intense feeling lasts anywhere from ten minutes to an hour or two at a time, and it is often accompanied by physical symptoms, such as sweating, palpitations, cold hands, and rapid heartbeat. Panic attacks rob you of self-confidence and self-esteem, which can have a negative effect on exam performance even if you are not having a panic attack at the time. If you experience panic attack symptoms, seek medical advice immediately, obtain medical documentation, and have the exam deferred.

2.3 Strategies when in the Exam

Exams typically come in two flavours: essay topics, and multiple choice questions, although some exams are a combination of the two. There are some useful strategies that you can use to maximize your performance when you are in the exam hall and actually taking either type of exam.

Essay Exams

Essay exams usually consist of two or three sets of essay topic options, grouped according to key areas of the course. You are required to pick one essay topic from each group. Sometimes, essay exams consist of a combination of long and short answer essay questions. Your own essay exams will likely be variations on the above. Here are some useful strategies:

- **Take time to read over all the essay topics**. Don't just rush into picking topics on the basis of a cursory glance through the exam sheet. Use ten minutes at the beginning of the exam to think about the exact wording of each essay topic. To test your understanding, essay options are sometimes written so that they express topics in a way that you are not immediately familiar with. Taking time at the beginning of the exam to carefully review all the options may help you recognize topics and make better choices

- **Circle keywords and key phrases**. When you have selected a topic and are preparing to write the essay, begin by paying close attention to the wording and circle keywords and phrases. Think about what they mean, their significance and how they relate to each other

- **Create an essay outline**. When you have paid attention to the wording and defined the topic in your own mind, think about the component parts of your essay. What are the main ideas or issues that need to be discussed? How should they be ordered? Where should your critical evaluation go, and what thesis will it support? Arrange the details of the organization of your essay in point form, or perhaps as a mind map. Taking time to write an essay outline helps you to define the topic in detail. The outline will act as a guide to help keep you on track as you write the essay. Writing an essay outline also has the advantage that if you do not complete the entire essay in the time available, the grader will be able to look at the outline and gain an impression of what you would have included. This may help your grade

- **Write on the topic as set**. One of the most common reasons for losing marks is writing on something other than the topic as set. Unless you remember nothing and are looking for part-marks, you need to ensure that your essay stays on track and addresses the topic as set by the exam. Pause from time to time to ensure that the essay is proceeding according to your essay outline. If it is not, either make the necessary corrections to what you have written, or reconsider the outline. There is nothing wrong with making changes as you proceed, as long as the essay does not deviate from the assigned topic.

Multiple Choice Exams

There is a myth that compared to essay exams, multiple choice exams are easy. The right answers are there in front of you. If you revise hard and memorize the facts, it will be easy to pick the right answers! The reality, however, is that multiple choice exams are no easier or harder than other types of exam. For multiple choice exams, although the right answers are there in front of you, there are typically many more questions that you have to deal with than in an essay exam. Moreover, just as with the wording of exam essay topics, multiple choice exam questions are often set up to test your understanding of a topic and not simply your memory of facts and details. Here are some strategies that may be useful when taking a multiple choice exam:

- **Don't start answering the questions right away.** Take ten minutes to browse through the exam and make a note of the question numbers for easier questions

- **Answer the easy questions first.** Don't waste time struggling over difficult questions. This will help make you feel more at ease and bolster your self-confidence

- **When you have answered the easier questions, go back and answer the more difficult ones.** Being more relaxed, and having exercised your understanding in answering the easier questions, you are now in a better frame of mind to tackle the tricky questions

- **Do not jump to conclusions before reading the entire question.** Multiple choice exams test your understanding, not just your memory of facts. So there may be more than one acceptable answer to the question, yet only one answer that is most appropriate. Make sure you have read and thought about <u>all</u> the options before making a decision

- **Circle or underline keywords and key phrases in questions.** Multiple choice exams test your understanding of a topic in the way that the question is worded. So take note of important words and think about their meaning, significance, and how they relate to each other

- **Pay attention to words like "ALL," "ALWAYS," "SOME," "FEW," "MANY," "NONE," "NEVER," and "SOMETIMES."** These are known as "modal" words and they radically influence the meaning of a question. If you mistakenly read an "all" as a "some," for example, this could lead you to misunderstand the question and pick an inappropriate option or not know how to answer the question at all

- **Consider answering questions on the same topic areas together.** Working on questions from the same topic areas can be helpful because the questions as a group may offer clues about which are the best options. Be careful in recording your answers on the computer-readable scantron sheet, if there is one

- **Be prepared to change your answer**. Nothing is written in stone, and as you work through the exam and exercise your memory and understanding, it is inevitable that you will come to realize that you answered an earlier question wrongly. This does not mean that you have answered all your questions wrongly. It is perfectly normal to correct your answer to a question or two and nothing to get stressed out about

- **Remember that you are looking for the best answer, not merely a correct answer**. Multiple choice exams use *distracter options*. These may be correct answers to the question, but not the most appropriate answer. You need to read and think carefully about all the options in a question before making a choice. Watch out for modal words (see above)

- **Length distracters**. Sometimes, a multiple choice question is set up so that one option stands out from the rest by being very short or very long. Be wary of these. The brain is designed to notice differences, and there is a natural tendency to go for the option that stands out from the rest. The option that stands out from the rest because of its length may be the most appropriate answer, but it might equally well be a length distracter, designed to catch those who do not really understand the topic the question is asking about!

2.4 Other Exam Issues

On the day of an exam, the frenzy and excitement of it all can lead to one forgetting some of the more mundane, but nevertheless important things connected to the exam itself. Make sure that you know where the exam is taking place and at what time. Take with you all the required ID documentation, as invigilators will check the identity of students. Many exam halls have notices saying that food and drink are not allowed. But going without anything to eat or drink during a three-hour exam can create problems. There is nothing more distracting to a student than to be thirsty, or to have a rumbling, gnawing tummy as s/he tries to concentrate on the exam. Take some bottled water and a snack with you. Obviously, you should not bring a three-course dinner! Take something that you can eat surreptitiously as you work on the exam – maybe a candy bar, salad or a piece of fruit. Finally, for emotional comfort take with you a small trinket that carries comforting meaning for you and which is small enough for you to be able to place on the desk as you work – maybe it's a key ring, a tiny teddy, or a lucky charm. Whatever works for you!

3. TIME-MANAGEMENT SKILLS

3.1 Creating a Term Schedule

To use time well, you first need to know how much of it you've got and when. You can work this out by creating a term schedule. A term schedule breaks each day of the week up into hourly units. Your activities and commitments (what they are and where they occur) are mapped onto these hourly units for each day of the week. As a university student it is likely that all of your repetitive activities and commitments (e.g. attending lectures, tutorials and labs) will occur at the same times each week for the entire term. Other repetitive activities, such as studying course readings and reviewing lecture notes can then be rendered stable by assigning them to the same free time-slots each week. In this way, you can create a predictable term schedule of activities and commitments as well as get into the habit of sustained, regular study of course materials and lecture notes. Here is a **fictitious,** simplified example of what a term schedule may look like before studying course readings and lecture notes have been added.

TIME / DAY	8:30-9:30	9:30-10:30	10:30-11:30	11:30-12:30	12:30-1:30	1:30-2:30	2:30-3:30	3:30-4:30	4:30-5:30	5:30-6:30	6:30-7:30	7:30-8:30	8:30-9:30	9:30-10:30
MONDAY	Kine1000 Lecture Comp Science Lecture hall A	Kine1020 Lecture Comp Science Lecture hall A		Kine1000 Tutorial 316 Tait		Biol 1010 Lecture Curtis Lecture hall J				6pm ⟷	Math1013 Lecture South Ross 105	⟷	9pm	
TUESDAY		Kine1020 Lab Stong 101N	⟶	⟵	Psych1010 Lecture Curtis Lec hall C	⟶					Volunteer work	⟷		
WEDNESDAY	Kine1000 Lecture Comp Science Lecture hall A	Kine1020 Lecture Comp Science Lecture hall A				Biol 1010 Lecture Curtis Lecture hall J		PKIN Gmnas I Main gym Tait	⟷					
THURSDAY				Math1013 Tutorial Petrie 103		Biol 1010 Tutorial Vari Hall Room 2113				6pm ⟷	Math1013 Lecture South Ross 105	⟷	9pm	
FRIDAY						Biol 1010 Lecture Curtis Lecture hall J								
SATURDAY		9am ⟶	⟶	⟶	⟶	⟶	PART-TIME JOB	⟶						
SUNDAY		9am ⟶	⟶	⟶	⟶	⟶	PART-TIME JOB	⟶						

The term schedule provides an easy to understand plan of your repetitive academic commitments. Pretend that the above example is *your* term schedule for a moment. You can see where your available time is each day. It is represented by the blank units. Adding up the blanks, this schedule shows that you have about fifty-five hours of time available to you each week (commuting time, lunch and dinner times etc. have not been included in the example for the sake of simplicity, but these reduce the amount of available time and you will need to take them into account when you create your term schedule for real). **You need to use some of this available time to schedule regular study times for course readings and for going over lecture notes each week throughout the term.** The amount of time needed for study can vary from person to person, but a total of approximately twenty hours of study time per week should be planned for at first, and adjust this as needed.

Where should you place study times in the term schedule? Notice that not all units of time are created equal! For example, how fresh and ready for study do you think you would be after eight and a half hours of work on Saturdays and Sundays, or after 9pm on Mondays having spent twelve and a half hours on campus?! **One secret to good time management is not to regard scattered hours between classes as being of no use except as coffee breaks - main points from an assigned reading can be revised even in a half hour!** So use the gaps between classes for study rather than waiting until the end of the day when you are tired. You should also avoid being fooled by the notion that because a particular day offers a lot of available time (e.g. Friday), all your study can be saved up and scheduled on that day. The human brain is quickly fatigued by mental effort. You will achieve more by studying for shorter periods but more frequently over different days of the week than by scheduling huge chunks of study time on the same day. Hence, the frequent gaps between classes are ideally suited to study.

Easing yourself into a study routine is important, but it is also important to build some flexibility into your term schedule. For each day of the week, try to keep at least one or two hourly units of time free. When nothing prevents you from studying at the scheduled times, use these free units of time to chill out and relax. When something unexpected prevents you from studying at a scheduled time on a particular day, simply transfer your study to one or more of the free hourly units of time that you have built into that day.

When you have placed your study times and free times in your term schedule, you need to think about *where* you will do your studying. All university campuses offer places for quiet work in libraries and computer labs. Some campus libraries offer study rooms that you can book in advance. So make an effort to ask around and find out what is available. **Don't assume that noisy campus is not the place for study and that all study should be done at home. This is a serious mistake** for the above reasons. In fact, for some people, noise is essential for study! You may be one of those people who need sensory stimulation (e.g. listening to music through your ipod) while you study. If you seem to study better this way, go with it. But if you are in a quiet zone (e.g. in a library), avoid creating noise for others - turn the volume down. Nuff said!

3.2 Creating a Monthly Schedule

A monthly schedule is a useful way of keeping track of non-repetitive activities and commitments that will come up in the not too distant future. These will not be indicated on your term schedule. Due dates for written assignments, as well as the dates of tests and exams are obvious candidates for your monthly schedule. A monthly schedule enables you to tell at a glance what your commitments are for any given month in the term. Pretend that the **fictitious** example below is from your own monthly schedule. You can see at a glance that there are two graded

---

OCTOBER

Written assignment #2 due on 16th in class for KINE1000

Lab quiz on 17th for KINE1020

Dental appointment on 31st

---

pieces of work coming up in October, one for KINE1000 and one for KINE1020. Notice that the times and location specifics are not given. That information is on your term schedule so all you have to do is cross-check these entries on your monthly schedule with information on your term schedule. Similarly, you don't need to specify the time and location of your dental appointment in your monthly schedule, as the time of the appointment will be in your weekly schedule and the location will no doubt be something you don't need to write down at all. It is important to keep entries in your monthly schedule as free from unnecessary details as possible, for the sake of clarity. Hence, duplicated information from your term schedule (e.g. specific location of the KINE1000 class – Computer Science lecture hall A) should not appear here. Instead, you would cross-check the entry in your monthly schedule with your term schedule. Your monthly schedule tells you *what* is coming up in the not too distant future, your term schedule tells you *where* (unless your professor advises of a different location), and your weekly schedule tells you *when* in the immediate future. Hopefully, you can see that your term schedule, monthly schedule, and weekly schedule work together to enable you to cross-check information about your commitments without unnecessarily duplicating the details and creating information clutter.

3.3 Creating a Weekly Schedule

A weekly schedule is essential for keeping track of non-repetitive commitments and activities coming up in the immediate future. For example, you don't need to write down each week that you have a KINE1000 lecture on Monday at 8:30am. You can keep track of these repetitive commitments simply by consulting your term schedule. But you do need to write down details of the readings assigned for lectures each week. You will also need to indicate which units of study time (from your term schedule) will be used for reading them. Assignment due dates and test or exam dates and times should also be written in your weekly schedule. For example, you need to write down that assignment #2 for KINE1000 is due on October 16th in class. In your weekly schedule, you also need to map the specific steps for completing assignments onto available units of study time from your term schedule. Take a look at the **fictitious** example of one day of the week from a weekly schedule. Notice how the non-repetitive activities of the day (research for an

---

FRIDAY 15th SEPTEMBER

8:30 – 10:30am   **Do keyword searches of academic journals for assignment #2 KINE1000**

11:30 – 12:30pm   **Read chapter 3 of textbook for BIOL1010 lecture**

3:30 – 4:30pm   **Do keyword search of academic books for assignment #2 KINE1000**

7:30 – 9:30pm   **Review BIOL1010 lecture notes and chapter 3 of textbook**

---

assignment, reading, and a review of lecture notes and reading) have been mapped onto units of study time in your term schedule (assuming that you had specified these units of available time as study times when you created the term schedule). Notice that all of your repetitive and non-repetitive activities and commitments for Friday 15th September can be understood and tracked by consulting both your weekly schedule and your term schedule.

3.4 Creating Task Assessments for Assignments

Being able to produce task assessments is crucial, both to good time management in general and to creating effective weekly schedules in particular. A task assessment involves two main stages. Firstly, the assignment is broken up into individual steps that have to be taken to complete it. Secondly, the amount of time needed to complete each step is estimated. Both stages require a bit of practice. Take a look at the example of assignment #2 instructions below from a first year kinesiology course:

> Write a researched five-page critical essay on how historical factors covered by the course affect how we move, display, judge and/or label human bodies in contemporary society.

How many steps do you think this assignment involves? Perhaps more than you think! Students new to doing task assessments often suggest three or four steps, as follows:

1. **Review course material to clarify which historical factors have been covered**
2. **Research the topic**
3. **Write a draft of the essay**
4. **Spell-check to produce final draft**

Steps two and three in particular combine a number of steps that need to be separated, as follows:

1. **Review course material**
2. **Locate sources on the topic through keyword searching for books and journals**
3. **Scan located sources to select those most appropriate for the assignment**
4. **Read selected sources carefully to identify authors' theses and supporting evidence**
5. **Critically evaluate authors' theses and supporting evidence**
6. **Formulate or modify own thesis in light of critical reading of authors**
7. **Write a draft of the body of the essay**
8. **Write drafts of the introduction and conclusion**
9. **Insert in-text citations for sources at the places where they are discussed in your essay**
10. **Construct a matching references page**
11. **Check for clarity, organization, and effective support of thesis to create a second draft of the entire essay. Then check spelling, vocabulary etc. for final draft**

See how easily four steps become eleven steps! And even here, steps have been left out for lack of space. It is important to identify all relevant steps. This makes for better estimates of the amount of time needed for assignment completion, and it allows you to allocate time more effectively. So breaking assignments up into steps is definitely something worth practicing.

For illustrative purposes, let us suppose that the above eleven steps complete the first stage of the task assessment. Now we need to estimate the amount of time needed to complete each of the eleven steps. This takes practice and experience, something you will gain over time. It is easy to underestimate the time needed. Here is a tentative estimate of the time needed for each step.

TASK ASSESSMENT

TASK: Assignment #2 Research Essay KINE1000
DUE: October 16th

STEPS	ESTIMATED HOURS
1. Review course material	2
2. Locate sources on the topic through keyword searching	8
3. Scan located sources to select those most appropriate for the assignment	2
4. Read selected sources to identify authors' theses and supporting evidence	12
5. Critically evaluate authors' theses and evidence	12
6. Formulate or modify own thesis in light of critical evaluation of the sources	1.5
7. Write a first draft of the body of the essay	7
8. Write first drafts of the introduction and conclusion	3.5
9. Insert in-text citations into essay where appropriate	2
10. Construct a matching references page	2
11. Check for clarity, organization, spelling and vocabulary to create a second draft of the essay	2
TOTAL ESTIMATED HOURS REQUIRED FOR COMPLETION:	**54**

The above task assessment for assignment #2 indicates that approximately fifty-four hours are needed to complete the assignment properly. It only remains to look at the task assessment, term schedule, monthly schedule, and weekly schedule together, in order to allocate units of available time from the term schedule to assignment steps as per the estimated hours. If you look at the example of a day from a weekly schedule above in section 3.3, you will see that three out of the eight hours estimated for locating sources are scheduled, from 8:30-10:30am and from 3:30-4:30pm on Friday September 15th. Obviously, you would need to allocate and fix in your weekly schedule units of available time reflecting the estimated time required for each of the eleven steps.

Three final points are worth mentioning:

- Different people feel more confident about different steps. For example, some students may be confident about their research skills or critical evaluation skills, while other students may not be confident about their ability in these areas but may be confident about drafting the body of an essay. This can affect one's estimated time for completing particular steps. You may feel confident, but it is best to err on the side of caution and **estimate more time than you think you will need** until your confidence is repeatedly justified by the grades and comments you receive

- **It is wise to give yourself plenty of room for maneuver. Complete the task assessments for assignments and get started working on the steps as soon as possible**, bearing in mind your other commitments and any important information scheduled for a lecture or tutorial later in the term that may be relevant to assignments. Fifty-four hours is a lot of hours in the above example. Assignments you are given may require less or more time than this for completion. The earlier you start on them, the more the total estimated completion time can be spread out, and the more extra available time there will be if things don't work out as smoothly as expected. Some courses include assignment instructions in their course outlines, which are given out right at the start of a course. When this happens, take advantage of it and perform task assessments for the assignments as soon as you have some confidence about the course and its material

- There will often be times when you have more than one assignment to do in the same period of time. After having done the task assessments for them, you may decide that there is simply not enough available time to properly complete all the steps of each assignment. **In situations where multiple assignments compete for available time, it is wise to allocate time in favour of assignments which carry higher percentage marks.** For example, if you have two essay assignments competing for the same available study times and the first assignment is worth 10% while the second is worth 20%, allocate a bit more time for the second assignment. But one circumstance where this rule might not hold would be if you were doing badly in a course that you had to pass, making a good mark on a 10% essay for this course more important than an excellent mark on a 20% essay for a course in which you were doing splendidly. Good time management will help you to do splendidly in all your courses.

4. ACTIVE LISTENING SKILLS

4.1 Active Listening versus Passive Hearing

If you have not come across it before, the phrase "active listening" may seem odd, perhaps even comical. Everyone is used to the idea of an "active lifestyle" or of "being active." These ideas involve moving the limbs of our bodies, and there is nothing strange about that idea. But how can you be "active" with your ears?! How can you *do* anything with those static flaps of gristle on either side of your head? Some people can wiggle their ears, but that hardly counts as being 'active' with them! In fact, active listening has very little to do with the ears. We *hear* with our ears, but we *listen* with our minds. **Active listening involves using one's mind to contextualize the ideas one hears at lecture and tutorial while actually being there hearing them.** In other words, active listening is an on-the-spot preliminary attempt to understand the deeper meaning and significance of things said. Passive hearing, on the other hand, involves little mental effort. Even a zombie can do it. With passive hearing, nothing is done with the ideas as they are being heard, other than perhaps to write them down. There is very little difference between a passive hearer and a tape-recorder, except that the tape-recorder is more efficient at capturing what is said!

Furnished with this rough distinction between active listening and passive hearing, stop a moment to ask yourself why you go to lecture and what you hope to get out of lectures. (It is worth taking time to reflect on this, since a lot of your life as a student will be spent in lecture.) Two sorts of answers tend to be given when students are asked this question:

1) I attend lectures primarily to ensure that I don't miss information that is relevant to tests, assignments, and exams

2) I attend lectures in order to gain a deeper understanding of the weekly topics and to develop an overall understanding and appreciation of the course and its subject-matter.

The first answer is typical of passive hearers. Students with this mentality regard themselves as sponges that must soak up all or as much as possible of the information given in lecture. They regard all information given in lecture as being of roughly equal importance, hence the need to record as much of it as possible. Passive hearers will be frantically scribbling down information all during lecture, or they will tune out altogether, leaving the capture of information up to an electronic recording device. They may also tune out if the lecture is structured around a PowerPoint presentation in the mistaken assumption that the slides will capture and explain what is important. The overall assumption of passive hearers is that lecture information needs to be recorded somehow in order to be able to revise it and remember it for assignments and exams. Nothing else needs to be 'done' with the information. In short, for passive hearers, lectures serve no other purpose than to present potentially testable information that must be captured and memorized. It is ironic that **it is active listening and *not* passive hearing which facilities effective memorization of facts and details.**

4.2 How to become an Active Listener

Passive hearers make the mistake of assuming that understanding = remembering. This is not so. Your laptop or desktop computer remembers things far more efficiently than you do, yet your computer understands nothing. If understanding is not remembering, then what is it? The answer, roughly speaking, is that **understanding = noticing connections (or potential connections or important disconnections) between ideas.** When you think about it, this is what contextualization amounts to. When we put an idea in context, we are essentially noticing how it is connected to other ideas. At lecture, active listeners are continually attempting to contextualize presented ideas by asking themselves such questions as:

> How is this point relevant to the current topic?
> How does this example illustrate the issue?
> Doesn't this point contradict what was argued in the assigned reading?
> Why is the lecturer repeating this point again and again?
> How is this idea relevant to the idea presented earlier in lecture?
> How does this topic fit in with the overall theme of the course?

Understanding is a matter of degree, not all or nothing. We understand an idea better the more we think of connections between it and a hierarchy of nested contexts, as illustrated below. A true

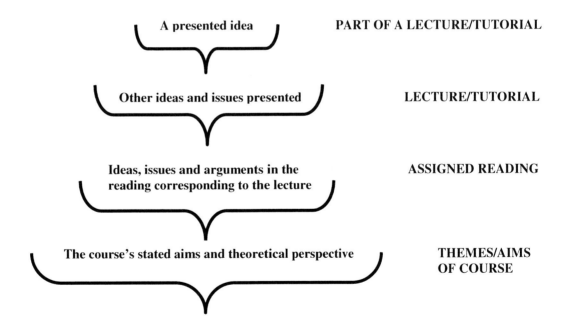

and deep understanding of a single idea presented in lecture depends on appreciating its connections with other ideas from lecture, on noticing the connections between ideas from lecture and the assigned reading, and on tracing connections between ideas in the assigned reading and the stated aims and background assumptions of the course. At university, assignments and exams generally test your understanding in this sense. They don't just test for how many facts you remember, although a by-product of active listening is that one does actually remember more and for longer than if one just tries to remember facts!

For all of the above reasons, passive hearing is not an effective way of connecting with lectures and tutorials. To shake off passive hearing habits and enter active listening mode, one needs to take the following steps:

1. Make the appropriate academic lifestyle choice. One has to choose to be an inquisitive and enthusiastic student who is interested in a course and its subject-matter for its own sake, not just for the sake of a grade. You need to foster this state of mind about all your courses in order not to slip into passive hearing mode

2. Take time to reflect on the ideas in the description of the course given out at the beginning of the course or available beforehand. Refer back to it again and again as the course progresses. For example, here is a description of a first-year kinesiology course:

> Students are introduced to the idea of the socially constructed body in the context of Western history and philosophy focusing on specific cultures and the social body in the discipline of sport sociology. That the body is socially constructed (as well as a biological organism) means that no human being lives outside of society. We all experience lifelong socialization as embodied persons interacting within specific social environments. An individual's social body is categorized and trained into socially-approved roles formulated by cultural imperatives and historical conditions that make particular impact around one's perceived gender, appearance, age, sexual orientation, 'race', ethnicity, ability, and class or caste. Studying the social body, we find that standards of health, performance, and appearance are established by a society's ideas of what is the 'norm', typically set by that society's dominant power groups. In this way, certain 'bodies' are privileged over others, especially in areas of extensive training such as sport. As the core course of sociology/history in our Kinesiology undergraduate degree, KINE 1000 focuses on historical and contemporary examples of trained bodies in sport, physical education, and other areas of physical culture. Understanding the social body is key to a critical approach to health and human rights.

Look for key theoretical concepts and assumptions, as well as for stated aims. In the above description, the theoretical concept of the **'socially constructed body'** is emphasized. So you would spend some time thinking about and researching what this actually means and what its implications are. Are there any opposed theoretical approaches (e.g. sociobiological) to explaining social phenomena? How do they contrast with the idea of the socially constructed body approach? Taking a little time to think about theoretical concepts of a course, their implications, and contrasting them with opposed approaches (if there are any) really helps one to gain a clear understanding of a course's basic assumptions. Turing now to course aims, a stated aim of the course is the bit that goes: **"Studying the social body, we find that standards of health, performance, and appearance are established by a society's ideas of what is the 'norm', typically set by that society's dominant power groups."** You would then think about this in relation to the rest of the course description and try to re-phrase this aim in your own words, for example: "Students should come to understand how and why inequalities and biases concerning gender, appearance etc. within the sports and physical education contexts do not reflect natural, biologically-based differences but instead, reflect the way society has been organized." You might then reasonably anticipate that much of the course and its assigned readings would be taken up with presenting and discussing issues and arguments illustrating the above. In summary, spending a little time to get clear on both theoretical concepts and stated aims of a course can pay huge dividends when it comes to actively listening at lectures and tutorials. You will find that you *can* trace connections between ideas and put them in context, because you have some idea of what the course and its readings are all about right from the start!

3. Do assigned weekly readings <u>before</u> lecture. If you don't do this, you cannot actively listen at the lecture with the reading in mind

4. At lecture, don't worry about getting all the information down. Focus on important ideas as they come up and spend time thinking about how they connect with the assigned reading and with the aims and assumptions of the course. Spend time at lecture thinking about this and expressing your thoughts about it in your lecture notes, even if the lecture is moving past the idea you are focusing on. Do this for as many main ideas from the lecture as you can, but don't fret about not covering everything. Leave that pleasure for the passive hearers!

5. Make a note of ideas you don't understand, disagree with, or about which you are very enthusiastic or interested. Attend your tutor's and professor's office hours to discuss them. Talking about ideas is the natural complement of active listening, and you will achieve a deeper understanding from discussing ideas in tutorial and office hours than passive hearers ever could, no matter how completely their lecture notes captured what was said at lecture

6. Do not arrive at lectures late. Often, the aims of a lecture are stated at the beginning, and this can be very helpful to understanding why a lecture focuses on particular ideas and issues. If you are even a few minutes late, you may miss this and have to work harder to understand the lecture than you otherwise would

7. Observe the idiosyncrasies of your professors, and take them into account at lecture. No two professors are exactly alike in their lecture styles. Professors may be very businesslike and have carefully set lecture plans that they follow rigidly right from the start of lecture, with little or no small talk. Some professors may operate with a loose lecture plan around which they improvise, perhaps with an informal introduction, the function of which is to say hello and connect with the audience. Other professors may not operate with lecture plans at all. Get to know the lecture styles of your professors, as this will help you to anticipate when important material is about to be presented and when you can take time to tune out a bit and think more about ideas that have already been presented.

5. NOTE-TAKING SKILLS

Good lecture notes are the product of active listening and they reflect the way active listening works. In other words, effective lecture notes facilitate the drawing of connections between ideas. Notes that simply capture everything said in lecture, perhaps in point form, do not reflect active listening and they are most likely the product of passive hearing. Such notes have a very short "shelf-life" and quickly become dead and meaningless. They are not very effective in helping one to remember anything or to develop one's understanding of ideas. Lecture notes that reflect active listening, in contrast, retain their meaningfulness beyond the lecture and function as a tool to help one remember and understand the deeper meaning and significance of ideas. Two styles of note-taking in particular can reflect active listening at lecture. These are Mind Map and Cornell styles of note-taking.

5.1 Mind Map Lecture Notes

Why is this style of note-taking called "Mind" Maps, you ask. Part of the answer is that mind maps are regarded as enabling one to make a visual record of one's non-linear thought-processes while actively listening to a lecture. Here is a fictitious excerpt of thought-processes going on in someone's head while actively listening (letters have been used in place of ideas):

> How does example B illustrate the lecture topic A? I can see that it is part of C. (**Later in the lecture**) ah, I see that C causes D. C causing D supports author's thesis E in last week's reading. (**Later in the lecture**) hmmm. Aha, I get it. I see now from E that the reading is all about F....... But what are the implications of F for A? Surely they are completely different issues? I get the feeling they are connected but I don't see how.

Here's how this active listening thought-process might look as represented by mind map lecture notes (again, letters have been used but in real life you would use keywords for the actual ideas):

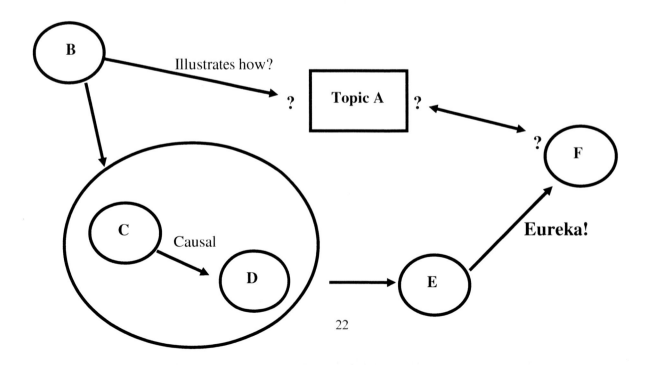

You can see from the example that the basic style of mind map lecture notes is quite simple. Key words and phrases are used for important ideas presented in lecture, and arrows represent your mind making connections between them as you actively listen to the lecture. Notice that you can use words or symbols to indicate what type of connection it is – for example, whether the connection is illustrative, causal, or conceptual. Notice also that question-marks can be used to indicate connections that you do not fully grasp or are critical of. **Including your failures to grasp ideas at lecture in your lecture notes is very important.** When you later come to look over your notes, being able to see what you didn't grasp within the context of what you did grasp guides you in your review of readings and course themes for the purpose of filling in gaps in your understanding. Mind maps make it very easy to represent connections not grasped within the context of connections grasped because they give you an easy to create and understand visual representation of your thinking about how lecture ideas are interconnected. For this reason, they can make an excellent after-lecture tool for developing one's understanding of a lecture and its related readings, themes and aims.

Another great feature of mind maps is that they are quicker and easier to create, add to, and amend than notes that capture lecture ideas in full sentences. If you make a connection between two ideas in a mind map and later in the lecture you realize that this was wrong, simply scribble out the connection and draw a new arrow at what you now take to be the right connection. This takes moments, while you would need to cross out a sentence and write another one if you were capturing lecture ideas in full sentences. This could take a long time if the point replacing the one crossed out is complex!

If you are not yet sold on the idea of using the mind map lecture-note style, here is another illustrative example of its effectiveness. Below is a small part of an imaginary first year kinesiology lecture. Imagine that the lecturer is speaking it and that you are a student in the audience. After this is an example of corresponding mind map style lecture-notes, followed by an example of how it might be filled in further after reviewing the initial mind map and relevant course readings.

> The concept of 'social determinants of health' challenges the traditional idea of health as solely within the skin of the individual. Traditionally in western thought, health has been regarded as the good working order of the various organs and sub-systems of the body. On this way of thinking, lack of health is inevitably conceptualized as caused by the breakdown in proper functioning of organs and sub-systems within (and including) the skins of biological bodies. Factors other than the purely biological, however, have increasingly come to be regarded as determining people's level of health. Some of these factors are social, hence the title of this lecture. Social factors that determine people's level of health include housing, educational level, conditions of employment, and income level. Astonishingly, the quality of factors such as these are better predictors of health than factors geared to the traditional ideas of health determiners, such as lifestyle choices (smoking, drinking, and exercise). Clearly, the concept of social determinants of health goes some way to shaking off dualistic models of health that have been with us for at least the last two-thousand years.

Mind Map lecture notes taken at lecture

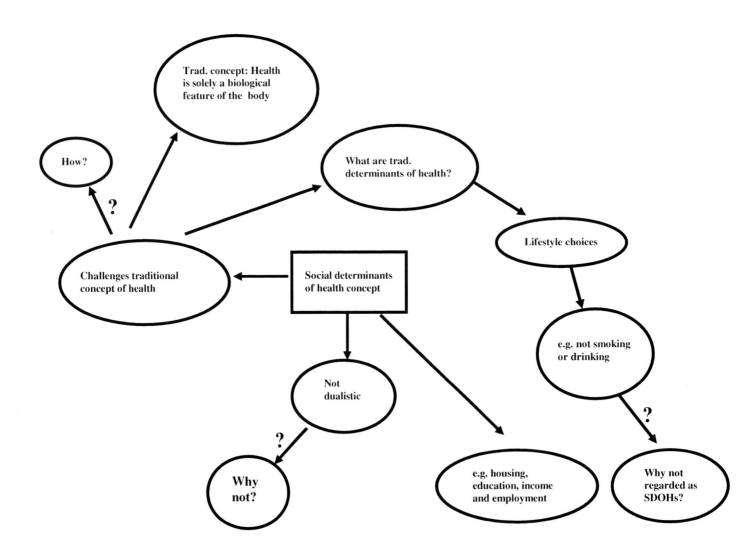

These mind map lecture notes reflect your mind actively making connections between ideas. The notes also show where connections failed to be made. In this way, the notes guide you as to what to focus your weekly review on when you look again at the assigned reading for this lecture. Below, the notes have been filled in (with dotted lines so that you can easily see what's been done) after the weekly review of these notes and the assigned readings.

Mind Map lecture notes filled in after reviewing assigned reading

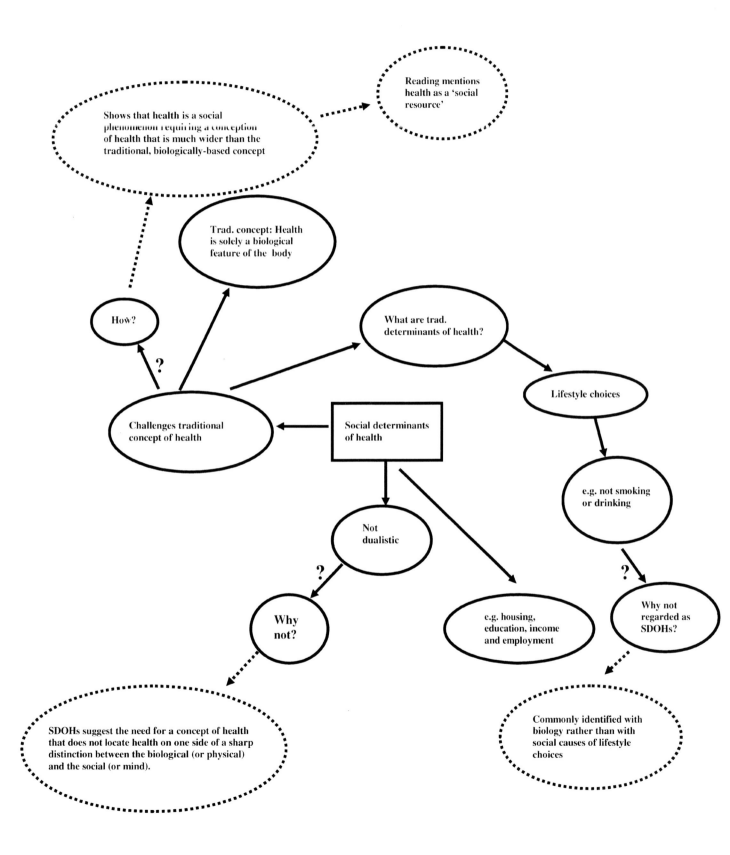

5.2 Cornell Style Lecture Notes

Why is this style of note-taking called "Cornell" notes? The answer is that it was first worked out by a professor of education, Walter Pauk, at Cornell University. The Cornell style of note-taking lends itself to a variety of adaptations and uses. Here we will discuss how it can be used to take notes about one's thought-processes as one actively listens at lecture.

Take a standard letter-size sheet of paper and draw a vertical line about 6cm from the left margin. Next, draw a horizontal line about 6cm from the bottom of the sheet. You will then have something that looks like this:

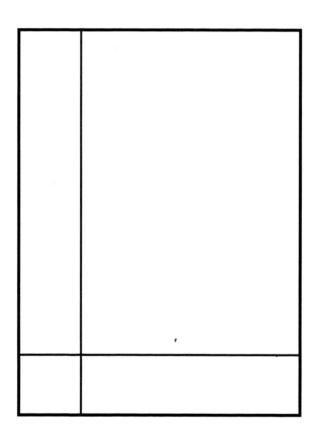

Use the wide column on the right to take down main ideas and useful supporting details from the lecture. Use the narrower column on the left to respond to ideas written in the right-hand column. Responses in the left column would include such things as summarizing main ideas from the right column, noting connections between ideas, and noting when you don't understand something about an idea in the right column. When recording not understanding something, you would leave a gap in the right column so that you can fill in details when you later review your lecture notes. Use the horizontal column at the bottom to summarize how the ideas in the right column connect with other readings, and with general themes and aims of the course. All of this is illustrated by the following example of Cornell style lecture notes that might have been made for our previous, imaginary excerpt from a lecture. The italics indicate places where the notes have been filled in after reviewing them later with the assigned reading.

Course: KINE1000 Lecture: Social Determinants of Health. Date: 15/9/04

ACTIVE RESPONSE COLUMN	LECTURE NOTES COLUMN
SDOHs challenge trad. concpt of health	Concept of social determinants of health (SDOHs) challenges trad. concept of health as within skin of individual.
HOW?	*Shows that health is a social phenomenon requiring a concept of health that is much wider than the traditional, biologically-based concept. Health is a resource (from assigned reading).*
Trad. concept: Health is biological ↓	Traditionally, health = proper functioning of body parts.
Poor health = biological causes	Poor health = body parts not working properly.
SDOHs are socio-political	Some SDOHs: Housing, education, employment, income.
SDOHs are better predictors of health	SDOHs ⟶ better predictors of health than trad. factors......e.g. smoking or drinking.
WHY SMOKING NOT REGARDED AS AN SDOH?	*Because it is commonly identified with biology without any acknowledgement of social determinants of lifestyle choices.*
SDOHs imply a non-dualist concept of health	The concept of SDOH shakes off dualistic models of health of last 2000+ years.
WHY?	*SDOHs suggest the need for a concept of health that does not locate health on one side of a sharp distinction between the biological (or physical) and the social (or mind). Also, SDOHs suggest that health is best understood as a group or class phenomenon determined by quality of resources. Dualism emphasizes that health is best understood as something that belongs strictly to <u>individuals</u>.*
SUMMARY OF CONNECTIONS (WITH OTHER READINGS/COURSE THEMES/AIMS) COLUMN:	*Concept of SDOHs fits in with the social constructionist assumption of the course: Level of health has more to do with the way society is organised than with biology. Differing levels of health across social groups and classes reflect unequal distribution of resources in society.*

There are some features worth noticing about the Cornell style lecture notes above:

- **Main ideas are recorded in a sequential way** that reflects the order of presentation of ideas in the lecture. It might be harder to take Cornell style notes in a lecture where the lecture style involves a lot of jumping from one topic to another. Mind mapping may be the best way to go in such cases. Similarly, where lectures present ideas in an ordered, sequential way, the Cornell style of note-taking might be a better approach

- **Sentences used to take down ideas while at the lecture are often shortened**, sometimes with the aid of symbols like the '=' symbol. Our use of this symbol in the above example would give mathematicians a seizure, but it is one useful symbol among others (e.g. arrows, dashes) for helping to capture ideas in shortened sentences. Other ways of shortening sentences include leaving out conjunctions and abbreviating long words

- **The active response column includes some mind map features.** We used an arrow symbol to indicate an inferential connection between two summarized ideas. Indeed, there is nothing wrong with this and the use of arrows, as in mind maps, can be a very useful way of quickly recording your thoughts about connections between ideas as you actively listen to the lecture

- **More has been written into the notes after lecture during your review than was written during the lecture itself! This really applies to mind map notes too.** It illustrates the importance you should attach to reviewing your lecture notes and associated readings. This is something that should be done, preferably on the same day as the lecture, but not so late in the day that you end up doing it when you are tired

- **The lecture notes indicate the course, the title of the lecture, and the date. Again, this is something that should apply to mind map notes too.** Even if you have labeled folders dedicated to different courses, sheets of paper can and do escape. Similarly, within a course folder sheets of paper can get out of sequence. If you put the course code, lecture title and date on each sheet of your lecture notes, you will always find it easy to know what course and lecture your notes are from

- **The lecture notes are personalized and mean most to the one who produced them. Once again, this applies to mind map notes as well.** Arrows, dashes, and abbreviations that mean something to someone when they write their lecture notes may not mean anything or may mean something different to someone else who reads them. Connections drawn between ideas reflect the personal understanding of the note-taker. They may be wrong, you may not understand them since you did not go through the process of actively working out the connections, or they may not agree with your understanding of the issue if you took the time to write and review your own notes. For these reasons, **always try to write your own notes rather than relying on those of another.** For the same reasons, you should always actively listen at lecture and write notes even when the lecture is based on a PowerPoint presentation, because PowerPoint presentations are in essence other people's lecture notes and not yours, in this case, the lecturer's.

6. ACTIVE READING SKILLS

When we first start out at university, we tend to bring ineffective reading habits with us, which are grouped together under the label "passive reading." Passive reading includes the following bad reading habits (some of them may be familiar to you!):

- Reading every word and sentence of the book or article, starting at the beginning and working one's way through to the end

- Assuming that everything in the book or article is of equal importance

- Assuming that the reading process involves a very careful, one-time reading of the book or article; you don't read the book or article more than once

- Not bothering to think about the different functions of different sections of the reading

- Not bothering to think about what type of reading it is, its function and target audience

- No bothering to think about how ideas in the reading hang together

- Not bothering to think about how the reading connects with other readings in the course

- Not bothering to think about how the reading fits in with the main themes and messages of the course.

Learning to read actively involves reversing the negative reading habits above. It is rather like learning to evaluate an ancient artifact. If you were asked to assess a strange object from an archeological dig, you would pick it up, handle it, turn it around in your hand noting all its features, and then after some reflection on its features, perhaps connecting them to what you know about other artifacts found at the site, and connecting them to what you know of the civilization that made this particular artifact, you would come to a judgment about what its function might have been. **With active reading, you examine each assigned reading for a course as if it is a strange artifact. Start by noting what kind of reading it is.**

6.1 Different Kinds of Course Readings

Is it a textbook? If it is, the contents will be clear and highly organized under headings and sub-headings – look to see if this is so. Its chapters or sections will likely cover all the topics focused on in the course. Look at the table of contents page at the front, and at the index at the back, and compare these to the course outline. The purpose of a textbook is to provide a clear, comprehensive overview of a subject-area, with a first or second-year undergraduate readership in mind. Usually, the chapters of a textbook are designed to be self-contained, which means that you can dip into a textbook at any chapter and make sense of the chapter without having to have read all the preceding ones. If your course has an assigned textbook, you have a ready-made clear overview of the entire course, and it would be worth looking over the chapters and sections and comparing them to the course outline and lecture schedule. Do this at the beginning of the course or even before the course starts, to get a feel for how the course will unfold and how the topics relate to each other.

Is it an academic book relating to some specific topic within the course? If so, its content may not be as highly structured as the content of a textbook. You may need to read earlier chapters to make sense of later chapters. Academic books have a wider target readership than textbooks, including undergraduates at all years of study, graduate students, and professors. Academic books tend to be less clear than textbooks (some of the active reading strategies later in this section will be helpful for reading academic books effectively).

Is it an academic journal article? Academic journal articles report the research of experts. They are written for other experts in the field, and for this reason they are often harder to understand than other kinds of readings. However, because academic journal articles report on cutting edge research, they tend to be argumentative, with a thesis supported by argument or evidence, and they are usually more structured than chapters in an academic book (some of the active reading strategies later in this section will be helpful for reading academic journal articles effectively).

Is it a course reader? Course readers often contain a selection of different kinds of readings by different authors, including book chapters, academic journal articles, and other types of academic sources. If there is a table of contents page, review this and the course outline together, to give you a sense of how the course will unfold. Approach each reading within the course reader according to the kind of reading it is. For example, if a particular reading in the course reader is a chapter or section from a textbook, look for clear descriptions and explanations of main ideas and principles. If it is a journal article, look for a thesis and supporting evidence.

6.2 Active Reading Techniques

With a background understanding of the kind of reading you are dealing with on any particular occasion, you are ready to go forward to actively scan the reading successively (i.e. more than once) to take note of the following specific features:

1) **Scan the title**. The title often provides useful, descriptive clues about the topic and focus of a reading. Look for any key concepts or ideas in the title. Reflect on what the title means and try to predict how the reading will unfold or what it will conclude

2) **Scan headings**. Headings throughout a reading are like signposts telling the reader where they are in the process of reading. Headings can be quite descriptive, and they can give you a good idea of what a reading is about. Write out the headings on a separate sheet of paper in the same order as they appear in the reading. Then reflect on them, together with the title. What do they suggest about the focus and direction of the reading?

3) **Scan the introduction and conclusion**. The reading may actually have headings entitled "Introduction," "Conclusion," or "Summary." But it is also possible that the reading contains no such headings. Whether or not there are such headings, the introduction comes at the beginning of a reading and the conclusion comes at the end. The beginning and end are important places because it is here that the author provides a short summary of the entire reading and indicates the thesis. So your next step in active scanning is to scan the introduction and conclusion of the reading, taking note of any summary, emphasized main ideas or issues, and indicated thesis. Look to see if the author offers a summary of the main pieces of evidence or reasoning and how they support the thesis. Highlight these places in the text and write them out on a separate sheet of paper, perhaps in mind map form. Reflect on the meaning and significance of the thesis, and look over the thesis and supporting details together with the title and headings that you have already located and written out

4) **Scan entire reading for main ideas and arguments**. With the above steps completed, you should have a good initial impression of what the reading is about, its focus, how some of the ideas hang together, and the thesis the author is defending. Now it is time to read through the entire reading quickly. Use your acquired initial understanding to distinguish and notice further main ideas and arguments in the reading. Use a highlighter to highlight these important places as you go through the reading

5) **Reflect on connections to other readings and to themes of the course**. Having completed the above quick scans, you should have a good working knowledge of the reading. Now take some time to reflect on how the thesis connects with main ideas of other readings, lectures, and with themes and messages of the course. Make a note of these connections, perhaps filling in some details as you review related Cornell or mind map lecture notes

6) **Scan the reading slowly and carefully.** The above steps have acted rather like a telescope. An object way in the distance is fuzzy and indistinct, and it is hard to tell what it is at first. But when you train your telescope on it and turn the adjustment, the object and its details begin to come into focus until you are able to see clearly what it is. Until you have performed the above active reading steps, a reading is as fuzzy and indistinct in its meaning and significance as an object in the distance. Going through the above steps is like turning the adjustment on the telescope until everything is clear and visible. Now that you have a good working understanding of the meaning and significance of the reading, you are ready to get much more out of it as you scan it slowly and carefully than a student would who neglected the above steps and started and finished with one slow, careful read. This is a mistake that must be avoided. It is like trying to describe and understand in detail an indistinct, fuzzy object in the distance without having used a telescope to bring it into focus!

6.3 Active Reading Exercise

The following is an abridged, version of a journal article by Emily Martin, first published in 1991 in *Signs: Journal of Women in Culture and Society*, *16*(31), 485-501. Endnote numbers and endnotes have been removed. Use this reading as an exercise in active reading techniques. Apply the techniques described in the previous section, and compare your overall reading experience with your prior experiences as a passive reader.

THE EGG AND THE SPERM: HOW SCIENCE HAS CONSTRUCTED A ROMANCE BASED ON STEREOTYPICAL MALE-FEMALE ROLES

[...]

As an anthropologist, I am intrigued by the possibility that culture shapes how biological scientists describe what they discover about the natural world. If this were so, we would be learning about more than the natural world in high school biology class; we would be learning about cultural beliefs and practices as if they were part of nature. In the course of my research I realized that the picture of egg and sperm drawn in popular as well as scientific accounts of reproductive biology relies on stereotypes central to our cultural definitions of male and female. The stereotypes imply not only that female biological processes are less worthy than their male counterparts but also that women are less worthy than men. Part of my goal in writing this article is to shine a bright light on the gender stereotypes hidden within the scientific language of biology. Exposed in such a light, I hope they will lose much of their power to harm us.

EGG AND SPERM: A SCIENTIFIC FAIRY TALE

At a fundamental level, all major scientific textbooks depict male and female reproductive organs as systems for the production of valuable substances, such as eggs and sperm. In the case of women, the monthly cycle is described as being designed to produce eggs and prepare a suitable place for them to be fertilized and grown – all to the end of making babies. But the enthusiasm ends there. By extolling the female cycle as a productive enterprise, menstruation must necessarily be viewed as a failure. Medical texts describe menstruation as the 'debris' of the uterine lining, the result of necrosis, or death of tissue. The descriptions imply that a system has gone awry, making products of no use, not to specification, unsalable, wasted, scrap. An illustration in a widely used medical text shows menstruation as a chaotic disintegration of form, complementing the texts that describe it as 'ceasing', 'dying', 'losing', 'denuding', 'expelling'.

Male reproductive physiology is evaluated quite differently. One of the texts that sees menstruation as failed production employs a sort of breathless prose when it describes the maturation of sperm: 'The mechanisms which guide the remarkable cellular transformation from spermatid to mature sperm remain uncertain....Perhaps the most amazing characteristic of spermatogenesis is its sheer magnitude: the normal human male may manufacture several hundred million sperm per day.' In the classic text *Medical Physiology*, edited by Vernon Mountcastle, the male/female, productive/destructive comparison is more explicit: 'Whereas the female *sheds* only a single gamete each month, the seminiferous tubules *produce* hundreds of millions of sperm each day' (emphasis mine). The female author of another text marvels at the length of the microscopic seminiferous tubules, which, if uncoiled and placed end to end, 'would span almost one-third of a mile!' She writes, 'In an adult male these structures produce millions of sperm cells each day.' Later she asks, 'How is this feat accomplished?' None of these texts expresses such intense enthusiasm for any female processes. It is surely no accident that the 'remarkable' process of making sperm involves precisely what, in the medical view, menstruation does not: production of something deemed valuable.

[...]

To avoid the negative connotations that some people associate with the female reproductive system, scientists could begin to describe male and female processes as homologous. They might credit females with 'producing' mature ova one at a time, as they're needed each month, and describe males as having to face problems of degenerating germ cells. This degeneration would occur throughout life among spermatogonia, the undifferentiated germ cells in the testes that are the long-lived, dormant precursors of sperm.

But the texts have an almost dogged insistence on casting female processes in a negative light. The texts celebrate sperm production because it is continuous from puberty to senescence, while they portray egg production as inferior because it is finished at birth. This makes the female seem unproductive, but some texts will also insist that it is she who is wasteful. In a section heading for *Molecular Biology of the Cell*, a best-selling text, we are told that 'Oogenesis is wasteful.' The text goes on to emphasize that of the seven million oogonia, or egg germ cells, in the female embryo, most degenerate in the ovary. Of those that do go on to become oocytes, or eggs, many also degenerate, so that at birth only two million eggs remain in the ovaries. Degeneration continues throughout a woman's life: by puberty 300 000 eggs remain, and only a few are present by menopause. 'During the 40 or so years of a woman's reproductive life, only 400 to 500 eggs will have been released', the authors write. 'All the rest will have degenerated. It is still a mystery why so many eggs are formed only to die in the ovaries.'

The real mystery is why the male's vast production of sperm is not seen as wasteful. Assuming that a man 'produces' 100 million (10^8) sperm per day (a conservative estimate) during an average reproductive life of sixty years, he would produce well over two trillion sperm in his lifetime. Assuming that a woman 'ripens' one egg per lunar month, or thirteen per year, over the course of her forty-year reproductive life, she would total five hundred eggs in her lifetime. But the word 'waste' implies an excess, too much produced. Assuming two or three offspring, for every baby a woman produces, she wastes only around two hundred eggs. For every baby a man produces, he wastes more than one trillion (10^{12}) sperm.

How is it that positive images are denied to the bodies of women? A look at language – in this case, scientific language – provides the first clue. Take the egg and the sperm. It is remarkable how 'femininely' the egg behaves and how 'masculinely' the sperm. The egg is seen as large and passive. It does not *move* or *journey*, but passively 'is transported', 'is swept', or even 'drifts' along the fallopian tube. In utter contrast, sperm are small, 'streamlined', and invariably active. They 'deliver' their genes to the egg, 'activate the developmental program of the egg', and have a 'velocity' that is often remarked upon. Their tales are 'strong' and efficiently powered. Together with the forces of ejaculation, they can 'propel the semen into the deepest recesses of the vagina'. For this they need 'energy', 'fuel', so that with a 'whiplashlike motion and strong lurches' they can 'burrow through the egg coat' and 'penetrate' it.

[...]

One depiction of sperm as weak and timid, instead of strong and powerful – the only such representation in western civilization, so far as I know – occurs in Woody Allen's movie *Everything You Always Wanted To Know About Sex**But Were afraid to Ask*. Allen, playing the part of an apprehensive sperm inside a man's testicles, is scared of the man's approaching orgasm. He is reluctant to launch himself into the darkness, afraid of contraceptive devices, afraid of winding up on the ceiling if the man masturbates.

The more common picture – egg as damsel in distress, shielded only by her sacred garments; sperm as heroic warrior to the rescue – cannot be proved to be dictated by the biology of these events. While the 'facts' of biology may not *always* be constructed in cultural terms, I would argue that in this case they are. The degree of metaphorical content in these descriptions, the extent to which differences between egg and sperm are emphasized, and the parallels between cultural stereotypes of male and female behavior, and the character of egg and sperm all point to this conclusion.

NEW RESEARCH, OLD IMAGERY

As new understandings of egg and sperm emerge, textbook gender imagery is being revised. But the new research, far from escaping the stereotypical representations of egg and sperm, simply replicates elements of textbook gender imagery in a different form. The persistence of this imagery calls to mind what Ludwik Fleck termed 'self-contained' nature of scientific thought. As he described it, 'the interaction between what is already known, what remains to be learned, and those who are to apprehend it, go to ensure harmony within the system. But at the same time they also preserve the harmony of illusions, which is quite secure within the confines of a given thought style.' We need to understand the way in which the cultural content in scientific descriptions changes as biological discoveries unfold, and whether that cultural content is solidly entrenched or easily changed.

In all of the texts quoted above, sperm are described as penetrating the egg, and specific substances on a sperm's head are described as binding to the egg. Recently, this description of events was rewritten in a biophysics lab at Johns Hopkins University – transforming the egg from the passive to the active party.

[...]

...[R]eseachers began to ask questions about the mechanical force of the sperm's tail....They discovered, to their great surprise, that the forward thrust of sperm is extremely weak, which contradicts the assumption that sperm are forceful penetrators. Rather than thrusting forward, the sperm's head was now seen to move mostly back and forth. The sideways motion of the sperm's tail makes the head move sideways with a force that is ten times stronger than its forward movement. So even if the overall force of the sperm were strong enough to mechanically break the zona, most of its force would be directed sideways rather than forward. In fact, its strongest tendency, by tenfold, is to escape by attempting to pry itself off the egg. Sperm, then, must be exceptionally efficient at *escaping* from any cell surface they contact. And the surface of the egg must be designed to trap the sperm and prevent their escape. Otherwise, few if any sperm would reach the egg.

[...]

SOCIAL IMPLICATIONS: THINKING BEYOND

These revisionist accounts of the egg and sperm cannot seem to escape the hierarchical imagery of older accounts. Even though each new account gives the egg a larger and more active role, taken together they bring into play another cultural stereotype: woman as a dangerous and aggressive threat. In the Johns Hopkins lab's revised model, the egg ends up as the female aggressor who 'captures and tethers' the sperm with her sticky zona, rather like a spider lying in wait in her web. The Schatten lab has the egg's nucleus 'interrupt' the sperm's dive with a 'sudden and swift' rush by which she 'clasps the sperm and guides its nucleus to the center.' Wassarman's description of the surface of the egg 'covered with thousands of plasma membrane-bound projections, called micro-villi' that reach out and clasp the sperm adds to the spiderlike imagery.

These images grant the egg an active role but at the cost of appearing disturbingly aggressive. Images of woman as dangerous and aggressive, the femme fatale who victimizes men, are widespread in Western literature and culture. More specific is the connection of spider imagery with the idea of an engulfing, devouring mother. New data did not lead scientists to eliminate gender stereotypes in their descriptions of egg and sperm. Instead, scientists simply began to describe egg and sperm in different, but no less damaging, terms.

[...]

The model that biologists use to describe their data can have important social effects. During the nineteenth century, the social and natural sciences strongly influenced each other: the social ideas of Malthus about how to avoid the natural increase of the poor inspired Darwin's *Origin of Species*. Once the *Origin* stood as a description of the natural world, complete with competition and market struggles, it could be reimported into social science as social Darwinism, in order to justify the social order of the time. What we are seeing now is similar: the importation of cultural ideas about passive females and heroic males into the 'personalities' of gametes. This amounts to the 'implanting of social imagery on representations of nature so as to lay a firm basis for reimporting exactly that same imagery as natural explanations of social phenomena.

Further research would show us exactly what social effects are being wrought from the biological imagery of egg and sperm. At the very least, the imagery keeps alive some of the hoariest old stereotypes about weak damsels in distress and their strong male rescuers. That these stereotypes are now being written in at the level of the *cell* constitutes a powerful move to make them seem so natural as to be beyond alteration.

The stereotypical imagery might also encourage people to imagine that what results from the interaction of egg and sperm – a fertilized egg – is the result of deliberate 'human' action at the cellular level. Whatever the intentions of the human couple, in this microscopic 'culture' a cellular 'bride' (or femme fatale) and a cellular 'groom' (her victim) make a cellular baby. Rosalind Petchesky points out that through visual representations such as sonograms, we are given 'images of younger and younger, and tinier and tinier, fetuses being "saved".' This leads to 'the point of visibility being "pushed back" *indefinitely*.' Endowing egg and sperm with intentional action, a key aspect of personhood in our culture, lays the foundation for the point of viability being pushed back to the moment of fertilization. This will likely lead to greater acceptance of technological developments and new forms of scrutiny and manipulation, for the benefit of these inner 'persons': court-ordered restrictions on a pregnant woman's activities in order to protect her fetus, fetal surgery, amniocentesis, and rescinding of abortion rights, to name but a few examples.

[...]

One clear feminist challenge is to wake up sleeping metaphors in science, particularly those involved in descriptions of the egg and the sperm. Although the literary convention is to call such metaphors 'dead', they are not so much dead as sleeping, hidden within the scientific content of texts – and all the more powerful for it. Waking up such metaphors, by becoming aware of when we are projecting cultural imagery onto what we study, will improve our ability to investigate and understand nature. Waking up such metaphors, by becoming aware of their implications, will rob them of their power to naturalize our social conventions about gender.

Below is a description of possible thought-processes that a seasoned user of active reading techniques might experience when actively reading the practice article above.

1) **Scan the title**. The title *The Egg and The Sperm: How Science Has Constructed A Romance Based On Stereotypical Male-Female Roles* offers some excellent clues about the topic of the article. A key phrase is "constructed a romance." A romance is a popular genre of novel – a work of fiction, with a plot usually based on a relationship between central male and female characters. Another key phrase is "based on stereotypical male-female roles." Stereotypical male and female roles include the ideas that the male is the more superior, dominant, powerful, active provider, while the woman is the less superior, passive, weak, stay at home, nurturing caregiver. The title, then, seems to suggest that the article is about the way science constructs a fictitious story about the relationship between the egg and sperm by projecting typical stereotypes of male and female onto the egg and sperm. Interesting!

2) **Scan headings**. Below, the headings and title have been written out on a sheet of paper.

> **The Egg And The Sperm: How Science Has Constructed A Romance Based On Stereotypical Male-Female Roles.**
>
> **Egg and Sperm: A Scientific Fairy Tale.**
>
> **New Research, Old Imagery.**
>
> **Social Implications: Thinking Beyond.**

The first section-heading seems to bear out the impression that the article as about the way science constructs a fictitious story about the egg and sperm. A key phrase here is "fairy tale." A fairly tale is a genre of extreme fiction – fantasy! This section-heading suggests that the fictitious story science constructs about the egg and sperm is described and explained in this section. The second section-heading is intriguing. Key terms here are "new" and "old." The heading seems to be suggesting that science has revised its understanding of egg and sperm, but that it still uses old imagery. What does "old imagery" mean here? Presumably, it means the same old story of stereotyped male and female roles. This section-heading also seems to vindicate the initial impression of what the article is about. The final section-heading is interesting. A key phrase here is "social implications." When we talk about 'social implications' of something, we are usually referring to negative social consequences. This section-heading suggests, then, that the fictitious story of egg and sperm that science constructs has negative social consequences. Summing up what we have learned so far, from scanning the title and headings, it would seem the article is about how science constructs a false account of egg and sperm by projecting male and female stereotypes onto them; even with new research about egg and sperm, the construction of this false account persists, and it has negative social consequences.

3) **Scan the introduction and conclusion**. The following main ideas and claims in the introduction and conclusion are worth highlighting and writing out on a separate sheet of paper, as below.

> **INTRO.**
> "...I am intrigued by the possibility that culture shapes how biological scientists describe what they discover about the natural world."
>
> "...I realized that the picture of egg and sperm...in scientific accounts of reproductive biology...relies on stereotypes...of male and female."
>
> "Part of my goal in writing this article is to shine a bright light on the gender stereotypes...within the scientific language of biology."
>
> "Exposed in such a light, I hope they will lose much of their power to harm us."
>
> **CONC.**
> "One clear feminist challenge is to wake up sleeping metaphors in science, particularly those involved in descriptions of the egg and the sperm."
>
> "...becoming aware of their implications, will rob them of their power to naturalize our social conventions about gender."

The first and second selected sentences from the introduction, as well as the first selected sentence from the conclusion clearly support the view that the article is about how science constructs a false account of egg and sperm by projecting male and female stereotypes onto them. The third selected sentence from the introduction is important. The phrase "part of my goal in writing this article..." indicates that a main thesis is about to be given. The phrase "shine a bright light on gender stereotypes" is a lively way of claiming that there *are* such stereotypes in the scientific language of biology and that this will be demonstrated. So, part of the author's main thesis would seem to be the claim that scientific accounts of egg and sperm are constructed with gender stereotypes. The last selected sentence from the introduction and the last selected sentence from the conclusion seem to be connected and illuminate each other. Taking into account what we have already discovered from the previous active reading steps, the last sentence from the introduction seems to mean that gender stereotypes in scientific accounts will lose their power to have negative social consequences if we become aware of their use. One obvious question is that of *how* they might have negative social consequences. We are given a clue by the last sentence of the conclusion, which states that they lose their power to "naturalize our social conventions about gender." Presumably, then, the author means that gender stereotypes in scientific accounts can have negative social consequences because they naturalize gender stereotypes present in society. But what does "naturalize" mean? It sounds rather like "natural." Reflecting on the first selected sentence from the introduction, "naturalize" might mean to make something seem like part of the natural world, in this case, what gender stereotypes say about male and female. But *why* would this have negative social consequences?

4) **Scan entire reading for main ideas and arguments**. Some important highlighted sentences as the text was scanned:

SCIENTIFIC FAIRY TALE

"The...common picture – egg as damsel in distress...sperm as heroic warrior to the rescue – cannot be proved to be dictated by the biology of these events. While the 'facts' of biology may not *always* be constructed in cultural terms, I would argue that in this case they are."

"The degree of metaphorical content in these descriptions, the extent to which differences between egg and sperm are emphasized, and the parallels between cultural stereotypes of male and female behaviour, and the character of egg and sperm all point to this conclusion."

NEW RESEARCH, OLD IMAGERY

"As new understandings of egg and sperm emerge, textbook gender imagery is being revised. But the new research, far from escaping stereotypical representations of egg and sperm, simply replicates elements of textbook gender imagery in a different form."

SOCIAL IMPLICATIONS: THINKING BEYOND

"The stereotypical imagery might encourage people to imagine that...a fertilized egg...is the result of deliberate 'human' action at the cellular level...a cellular 'bride'...and a cellular 'groom' make a cellular baby."

"Endowing egg and sperm with intentional action, a key aspect of personhood...lays the foundation for the point of viability being pushed back to the moment of fertilization. This will likely lead to greater acceptance of technological developments and new forms of scrutiny and manipulation, for the benefit of these inner 'persons': court-ordered restrictions on a pregnant woman's activities in order to protect her fetus, fetal surgery, amniocentesis, and rescinding of abortion rights, to name but a few examples."

Completing the prior steps of active reading made it easier to notice the important above-highlighted places in the text. The first highlighted sentence in the 'Fairy Tale' section contains an important phrase: "I would argue." This phrase can indicate a thesis. What is it that is argued? It is that the characteristics of egg (e.g. passive damsel in distress) and sperm (active, warrior to the rescue) common in scientific descriptions are not based on objective facts but on cultural gender stereotypes. The second highlighted sentence from this section reinforces this impression. It tells us that the author regards the results of her analysis of scientific literature as evidence for this thesis. The highlighted sentence in the 'New Research' section supports the earlier interpretation of this section-heading. The highlighted sentences from the 'Social Implications' section answer a question raised in the previous active reading step. Science is regarded as objective, unbiased, and about the facts of nature. Hence, the noted gender stereotypes in scientific literature of egg and sperm, which give egg and sperm stereotypically gendered personalities and abilities, render it more likely that the newly fertilized egg will be regarded, as a matter of "scientific fact", as a 'person'. This could threaten women's freedom to control their own bodies and erode hard-won legal rights safeguarding this freedom. These are the negative social consequences surmised on the basis of this section heading earlier in the active reading process. Summing up one's understanding of the reading so far: the author argues that scientific literature on the egg and sperm constructs its account of human reproduction using biased and unfair cultural gender stereotypes that work against women and cast them in a negative light – so science is not neutral and objective (this is a main thesis in the article). New research that discovers and acknowledges the active role of the egg still constructs its account of human reproduction with biased and unfair cultural gender stereotypes. Because science is regarded as neutral and objective, the *presence* of gender stereotypes in scientific literature 'naturalizes' them – making them seem like features of the natural world rather than like the cultural creations that they are. The stereotypes have the effect of making egg, sperm, and their products – fertilized eggs – seem like fully-formed 'persons' at the cellular level, and this may be taken as scientific 'fact' because of the naturalizing effect of science. This threatens women's freedoms and rights to control their own bodies.

5) **Reflect on connections to other readings and to themes of the course.** The course outline, given in section 4.2 of this manual emphasizes the 'socially constructed body'. This article illustrates how male and female bodies are socially constructed in scientific literature. It also illustrates how science, through naturalization, supports and helps to perpetuate unequal cultural constructions of gender, possibly at the expense of women's freedoms and rights. This latter point from the article connects with the course's goal of showing how study of the socially constructed body illuminates human rights issues. The article also provides an example of how health is socially constructed. Creating accounts of the egg and sperm using biased and unequal cultural gender stereotypes, female reproduction comes to be represented as wasteful and as involving the death of tissue 'necrosis' (unhealthy), while male reproduction is represented as vibrant and amazingly productive (healthy).

The last step of the active reading process is to scan the article slowly and carefully. It is the step that most passive readers start and finish with. With the benefit of a preliminary understanding of the meaning and significance of the article, developed over the course of the successive five steps of active reading above, you, as a trainee active reader have put yourself in a position to genuinely understand the article, its meaning and significance in depth, because you have brought the article into focus. Passive readers, with their one-step once only careful reading of the text will take longer to read the piece than it took you to actively read it, and they will get less out of the reading. What more do you need to convince you that you need to develop your active reading skills by practicing the above techniques on your assigned course readings!

7. RESEARCH SKILLS

All of your research will likely take place within the context of some sort of assignment, most often a written assignment such as an essay or report. Here, we will take a look at the skills involved in the general process of doing research for a written assignment. Effective research is crucial to the quality of a piece of written academic work. Ideally, more time should be spent on research than on writing the research up as an essay or report. In doing a research essay assignment, roughly 70% of one's time should be devoted to researching sources and 30% to writing the essay. Writing the essay is supposed to be the final, easy step after having done all the hard work (the research). Of course, in reality, for a variety of reasons a student may have difficulties with the writing stage. When this happens, one should seek support from a campus writing centre in order to avoid letting the time for writing expand and shrink the available time for research. Generally, grades given for research assignments go down as the 70% research time decreases and the 30% writing stage increases. Hence, well developed research and writing skills are of extreme importance for good grades. Your research skills will develop over time as you do a number of assignments involving research. Now let's take a look at the basic research steps.

7.1 Read and Understand Assignment Instructions

Assignment instruction sheets are extremely important documents that deserve to be studied very closely. A common mistake early on in the research process, however, is that students glance at them or scan them quickly without giving a lot of thought to the exact wording of the instructions or to the wording of the specified grading criteria. Another common mistake occurs when students take a rather relaxed attitude, assuming for example that an essay which just *describes* a particular issue will do even though the assignment instructions explicitly ask for a student's *own developed critical input*, with personal critical input clearly specified as a grading criterion! Pay careful attention to the wording of the following excerpt from an assignment instruction sheet.

Research Report: 20% of final grade **Due:** Monday February 26th in lecture
Length: 7 pages + title page + reference page
Format: 12 pt Times New Roman, one inch margins, double-spaced, numbered pages, no underline or boldface.
Grading criteria: 8 marks for thorough research and for appropriate organization of information according to the report structure; 8 marks for effective analysis of the topic, and for useful, justified recommendations; 4 marks for correct APA in-text citations and references.
Purpose of assignment: Professionals in kinesiology-related jobs are able to undertake researched analysis of issues/problems in order to create reports that offer informed, justified, and effective courses of action to help solve the problems. Assignment #4 emulates this professional activity.
Instructions: Choose a topic from the attached list. Pretend that you are a professional from within or outside of the organization that you are reporting to. Give yourself a title (e.g. 'Consultant to the Ontario Ministry of Education'). Put your title and organization on your title page (see the attached sample). Research the topic, clarify the issues, survey and evaluate main lines of thinking, summarize the topic and associated problems as you see them after having evaluated existing theories/evidence, and make recommendations to help resolve the issue. You are required to present this work in research report form. APA in-text citations and references required.

A cursory scan of these instructions could give a student the impression that they are being asked to write a seven-page critical essay with in-text citations and a references page, worth 20% and due on February 26th. A close inspection reveals that one is being asked to take on the role of professional researcher and to write a seven-page research report. The first obvious question to generate and reflect on, then, is:

- **What is a research report, and how does a research report differ from a research essay?**

If you were doing the assignment, you should want a crystal clear understanding of the answer to this question. In fact, part of the instructions "Purpose of assignment" gives you a clear statement of the function of a research report (in contrast to that of an essay). So part of the above question is answered by the instructions themselves if you read them carefully! But you would still need to find out how a research report differs in its organizational structure from the way an essay is organized, because it would be very important not to write your report using organization appropriate to an essay.

An examination of the instructions reveals that the 20% is broken up into three grading criteria:

    8 marks for thorough research and for organization of information in research report form;
    8 marks for effective analysis of the issue, and for useful, justified recommendations;
    4 marks for correct APA in-text citations and references.

You would need to pay careful attention to the exact wording of these grading criteria and spend time getting clear on what they mean. For the first one, you would want to generate and answer the following questions:

- **What is "thorough research"? (This section of the manual will help answer this question!)**
- **What is the organizational form of a research report?**

For the second grading criterion, you would want to generate and answer the following questions:

- **What is "effective analysis" of the issue?**
- **What does it mean for recommendations to be "useful"?**
- **What does it mean for recommendations to be "justified"?**

Without having spent time to think about and answer these questions, you would not really know what you were doing as you tried to complete the assignment. In addition to the grading criteria, look for any specific instructions. **If there are any, break them up into steps and try to connect the steps to the grading criteria. Think about the wording in each step.** In the example above, this is rather easy since the instructions already break the assignment up into steps for you. There appear to be nine main steps. Here they are, along with the sorts of observations and questions about them that you would want to generate and answer (in bold), as well as their connection to the grading criteria (in italics):

1) Choose a topic from the list.
   - **So I cannot choose a topic of my own? (Check with professor or TA if you want to do a topic of your own – don't just assume that you can do your own thing.)**

2) Pretend that you are a professional from within or outside of the organization that you are reporting to. Give yourself a title (e.g. 'Consultant to the Ontario Ministry of Education'). Put your title and organization on your title page (see attached sample).
   - **<u>Why</u> is the assignment asking me to pretend to be a professional?**

3) Research the topic.
   - **Does "researching the topic" differ from the other items on the list, such as 'clarifying the issue'? Or does researching the topic <u>consist of</u> these other steps?** *Whatever it means, this step clearly connects with the "8 marks for thorough research..."*

4) Clarify the issues.
   - **What does this mean? Presumably, "clarify" does not mean "evaluate" since evaluation is given as a separate requirement in the instructions. Does "clarify the issues" mean working out how many distinct issues there are and giving a clear description of them?** *Connects with the "8 marks for effective analysis of the topic..."*

5) Survey and evaluate main lines of thinking.
   - **What does "main lines of thinking" mean? How do 'main lines of thinking' differ from 'issues'? What does it mean to "survey and evaluate" them?** *This step also seems to connect with the "8 marks for effective analysis of the topic..."*

6) Summarize the topic and associated problems as you see them after having evaluated existing theories/evidence.
   - **What does "summarize" mean? Why am I being asked to repeat what the topic and problems are? How does "as you see them" fit in with the other steps of the assignment? What is the connection between evaluating arguments (i.e. theories and evidence) and my own perspective?** *Again, this step seems to connect with the "8 marks for effective analysis of the topic..."*

7) Make recommendations to help resolve the issue.
   - **Presumably "issue" here means <u>problems</u> associated with some of the issues? I note that "help resolve" means that I am not expected to come up with total solutions.** *Connects with "8 marks for useful, justified recommendations."*

8) You are required to present this work in research report form.
   - **What is the organizational form of a research report?** *Connects with "8 marks for organization of information in research report form."*

9) APA in-text citations and references required.
   - **What are in-text citations and references, and what is APA style?** *Connects with "4 marks for correct APA in-text citations and references."*

In summary, coming to understand a set of assignment instructions can be like a mini assignment in itself! But it is essential to have: a) a full understanding of the grading criteria, b) how they apply to steps in the assignment, and c) what those steps are. Without this knowledge, you don't really know what you are doing. So do take time to think carefully about these things whenever you are given an assignment.

7.2 Define the Topic

For a written assignment there will be a topic you have to write about. It is important to define the topic – to gain a clear understanding of what subject areas the topic connects with, and just as importantly, what it does not cover. Putting it simply, you need some idea about where the topic begins and ends. Without this, your research would lack direction and focus. This point applies, irrespective of whether the topic is provided as part of the assignment instructions or whether it is something that you are required to come up with yourself.

To define the topic, you must take some time to think about and identify the main subject areas connected with it. Let's take an example. Suppose that you are required to write a critical research essay and that the topic given to you or which you come up with is the following:

> Should physical education be mandatory at all high school grades?

How many main subject areas do you think are connected to this topic? You might be forgiven for supposing that there are just two: 1) Health, and 2) the current state of physical education within high school education. You would then most likely research the connection between physical exercise and health, and then write an essay defending the thesis that physical education

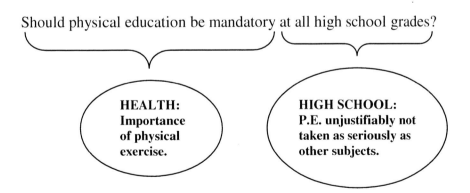

should be mandatory at all high school grades because of the importance of physical exercise for good levels of health. Your research would probably lead you to other relevant factors in defending your thesis, such as the importance of sport for developing both teamwork skills and personal character, and maybe even the point about a more healthy population being less of a burden on the public healthcare system. The problem here is that your thesis and related research would not have tapped into the controversial complexities of the issue. When the complexities and controversial aspects of an issue are not noticed or covered in an essay, it is sometimes because the original topic has not been thoroughly defined at the outset. Take another look at the sample topic. There is clearly a third subject area for research, namely the issue of mandatory education, of which mandatory physical education would be a particular example. When you think about it, how could you possibly address the issue of mandatory physical education without researching what educators think about the idea of mandatory education! Indeed, the fact that other high school subjects are mandatory is not of itself a justification for the claim that they *should* be. How much do *you* remember about the high school subjects you studied?! The point

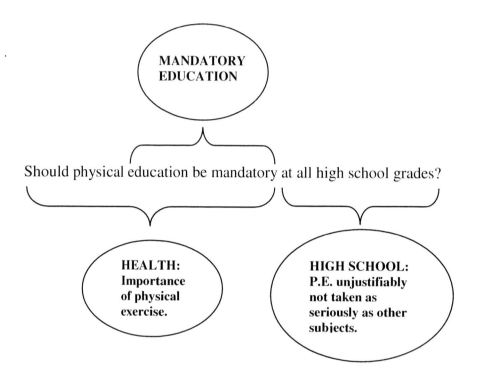

here is that when you define the topic thoroughly, its complexities and controversial dimensions become clearer as proper subject areas for research, leading to a more interesting and sophisticated essay. For example, along with the issues of physical exercise and health, and unequal treatment of physical education compared to traditional "academic" subjects at high school, you would also want to consider the question of mandatory education. Is it counter-productive? Might it inadvertently underscore the mistaken idea that certain body-shapes are more healthy and desirable than others? What is a 'healthy character' anyway? Could mandatory physical education at high school indirectly lead to the prevalent idea in society that those who stop exercising physically after graduation are less worthy of healthcare than others? Once you define the topic thoroughly, the more likely it is that all sorts of interesting ideas and questions will come flooding into your mind.

In addition to not taking enough time to reflect on the topic, students sometimes do not define the topic thoroughly because they are afraid to do so. They have limited time in which to complete the assignment, and a less thoroughly defined topic gives more of a sense that the issue is clear-cut and straightforward. **Clear-cut, straightforward issues are more quickly and easily dealt with than complex, controversial ones. You should avoid this way of thinking if you wish to achieve better grades for your work.** University is precisely about complex, controversial issues and ideas, and your professors and TAs want to see you grappling and struggling with difficult issues rather than reducing them to simple, straightforward matters. A good rule to keep in mind when presented with a written assignment is that nothing is clear-cut and straightforward if you take the trouble to think about it deeply enough.

Sometimes you may be given an assignment that requires you to come up with a research topic yourself. Here, **you need to be careful to find a topic that does not yield so many subject areas when you define it that it would be impossible to research them all adequately in the time available.** For example, suppose you came up with the following research topic: Violence in sport. If you defined this topic you would discover that it is connected to at least the following subject areas: Violence in soccer, violence in ice-hockey, violence in American football, violence in boxing, player violence, fan violence, physical violence, psychological abuse, and physical and psychological violence committed by coaches. All this would likely be far too much to research adequately in the time available. And trying to squeeze it all into an essay would undoubtedly lead to an unclear piece of written work lacking focus and direction. So be careful to select a manageable topic. For instance, instead of violence in sport, you might focus on a particular sport or category of violence, and have violence in soccer, or fan violence as your research topic. But remember, "manageable" should not be understood to mean "clear-cut and straightforward!"

7.3 Do Background Research

Students sometimes ask why background research isn't just lumped in with directly researching the topic for the assignment. Why is it a distinct stage? There are at least two reasons for why background research is a distinct step in the research process.

Firstly, background research is closely connected to the previous step. Indeed, background research can help you to define the topic. For example, you might have incompletely defined the topic: Should physical education be mandatory at all high school grades, only making the connection to the issue of health, and to the issue of the state of physical education in high school. In the course of doing background research in these areas, however, you would likely come across something on the issue of mandatory education, which could then lead you to realize that this is a third important subject area connected to the topic. **Doing background research is a way of developing a broad impression of the topic by becoming acquainted with general information from the subject areas that define it.** Defining the topic and doing background research work together in a 'boot-strapping' process: You need to define the topic in order to have some idea about how to focus your background research, and the background research helps you to define the topic further, leading you to refocus your background research.

Secondly, background research is a distinct stage in the research process because the sources from which you obtain information would include types not usually appropriate to use, cite or reference when writing the paper. Background research involves casting your net widely and making use of all sorts of sources in order to gain an initial impression of the topic that can guide you in further research for suitable sources. Indeed, you would *need* to do further research to locate suitable sources for the paper after you had done your background research. Students sometimes make the mistake of thinking that they can dispense with background research and go straight to locating suitable sources for writing the paper. If you are chasing better grades, you need to avoid this way of thinking. It is unwise to leave out background research, because it leaves open the possibility that you have not properly defined the topic, which nearly always results in faulty research and poorly focused written work.

Research via the Internet

How should you kick-start your background research? Perhaps the simplest way is to begin with some Internet research. To do this, you will need to use Internet search engines. Putting it at its simplest, an Internet search engine gathers information according to predetermined categories from the Internet and stores it according to predetermined categories. Hence, when you use a search engine, you are, in effect, searching a database defined by the parameters of the search engine. **What you are not doing is searching every word or phrase on every page of the World-Wide Web!** In other words, the World-Wide Web is not itself a database but a sea of information out of which databases are carved. Failing to realize this is a chief reason for why students don't find what they are looking for or only find limited, rather poor information when using the Internet. Different search engines collect different information and categorize it in different ways. What you fail to discover using particular keywords in one search engine you might discover using the same keywords in another. So it is useful to use a variety of search engines. The most popular ones include:

Google at http://www.google.com

AOL at http://www.aol.com

Yahoo at http://www.yahoo.com

Ask.com at http://www.ask.com

Internet Explorer at http://www.microsoft.com/windows/ie

You should check with your university library, as it may subscribe to some or all of these search engines and more, enabling you to use them without paying subscription fees. Check with your library to see how you can conduct Internet research remotely using your home computer but under the university library's agreements with these companies. This will usually involve creating a library account.

Use keywords and key phrases to search an engine's database. But what should you search for when doing background Internet research? No doubt you will key in very specific keywords and phrases to try to locate sources directly relevant to your topic. This is hard to resist – we all do it! You may get lucky and find some items of interest this way. The vast majority of relevant sources, however, will likely be invisible to you while you adopt this approach. **The most effective strategy is to search for reference sources that are more narrowly focused on subject areas relevant to your topic.** Many of these reference sources will have dedicated search engines of their own, enabling you to search every word or phrase within a source's database to locate relevant items that you would not be able to find just by using your general Internet search engine.

Reference sources to search for when doing background Internet research would include online indexes and bibliographies, encyclopedias, dictionaries, E-books, and Web Guides. You can find such sources by actually keying in words and phrases like these along with keywords for the subject area you want to investigate. Below is a list of some online reference sources that may be of use to you for background research. Your university library probably subscribes to some or all of them. Check your library's e-resources to find out.

Indexes and Bibliographies

General Science Index: Allows you to keyword search academic and general interest science journals and magazines on a wide range of topics.

Expanded Academic: Allows you to keyword search scholarly journals on a wide variety of topics.

Humanities International Index: Allows you to keyword search academic journals and general interest periodicals on a wide range of topics.

Canadian Periodicals Index: Allows you to keyword search academic journals and general interest newspapers and magazines, with an emphasis on Canadian issues.

Readers' Guide Abstracts: Allows you to keyword search general interest periodicals published in Canada and the United States.

CBCA Complete (Canadian Index): Allows you to keyword search abstracts related to current affairs and public policy in Canada.

Book Review Plus: Allows you to keyword search reviews of books and periodicals on a wide range of academic topics.

Factiva: Allows you to keyword search current and archival material from major international newspapers and magazines.

Encyclopedias

International encyclopedia of the social & behavioral sciences.

The Canadian encyclopedia.

Oxford reference online.

Encyclopedia Britannica online.

E-Books

Books@Ovid: Access to full texts of electronic handbooks on topics including clinical procedure, drugs and medication, diagnostics, and nursing.

Web Guides

The REHABDATA Database: Allows you to keyword search a database on disability and rehabilitation, including physical and psychiatric disability and related issues.

Pedro Physiotherapy Evidence Database: Gives you access to abstracts of rated, evidence-based clinical procedures in physiotherapy.

Primal Pictures: An animated, interactive model of the body allows you to explore the structures and functions of human anatomy.

Caution with Internet Research

You should take care when looking at information you have found via the Internet. Anyone can create a web site and disseminate information (including false and misleading information) on anything. Sometimes the communication of erroneous information is innocent, as when someone creates a web site or blog to express one's beliefs. Sometimes it is not innocent, as when beliefs are deliberately dressed up as facts to sway you politically, or when the information has a commercial bias for the purpose of encouraging you to buy a product. When confronted with information, always ask yourself about the source of that information. Is the author identified? Is the author affiliated with a university, government department, or with a respectable public or private institute or organization? Does the information itself come from a commercial or educational site, or from an organization or institute? One way to make a quick initial judgment is to look at the ending of the URL of the Internet site containing the information in question. If the URL ends with "**.com**" then the site is commercial. If it ends with "**.org**" then the site belongs to some sort of organization. If the URL ends with "**.edu**" then the site is educational.

Researching at the Library

A main attraction of Internet research is that it can be done from the comfort of one's own home, or anywhere. But not all reference sources will be accessible online. There will come a point in your background research when you need to be physically present at the library. For example, although you may be able to read abstracts of relevant sources via some of the online bibliographies and indexes listed above, you may still need to go to the library and locate the physically printed sources on the library shelves in order to read the full texts. Another obvious reason for why you need it be physically present at the library is that as part of your background research you need to look at relevant, physically published books and reference works in your library's catalogue. You may be able to search the catalogue remotely, but you will need to go physically to the library shelves to look at the sources. Pay special attention to the references section of your library. It is in this section that reference sources are located, including helpful dictionaries, bibliographies, indexes, and encyclopedias. Below is a list of some background research sources that you might find useful. Note that these sources have to be checked physically at the library.

Jenkins, Simon P. R. (2005). *Sports science handbook: The essential guide to kinesiology, sport and exercise science*. Brentwood, Essex: Multi-Science. **Library call number: G V558 J46.**

Stedman, Thomas L. (2006). *Stedman's orthopaedic & rehab words: Includes chiropractic, occupational therapy, physical therapy, podiatric, & sports medicine* (5th ed.). Baltimore, Md.: Lippincott Williams & Wilkins. **Library call number: RD 723 S74.**

Bartlett, Roger, Gratton, Chris, & Rolf, Christer G. (Eds.). (2006). *Encyclopedia of international sports studies*. London; New York: Routledge. **Library call number: GV 567 E486.**

Kent, M. (1998). *The Oxford dictionary of sports science and medicine* (2nd ed.). Oxford; New York: OUP. **Library call number: RC 1206 O94.**

Ostrow, Andrew C. (Ed.). (1990). *Dictionary of psychological tests in the sport and exercise sciences* (2nd ed.). Morgantown, W. Va.: Fitness Information Technology. **Library call number: GV 706.4 D57.**

Remley, Mary L. (1991). *Women in sport: An annotated bibliography and resource guide, 1900-1990*. Boston, Mass.: G. K. Hall. **Library call number: Z 7963 S6 R45.**

Shoebridge, M. (1987). *Women in sport: A select bibliography*. London: New York: Mansell. **Library call number: Z 7963 S6 S56.**

Arlott, J. (Ed.). (1975). *The Oxford companion to world sports & games*. London; New York: OUP. **Library call number: G V207 O93.**

Zeigler, Earle F. (1971). *Research in the history, philosophy, and international aspects of physical education and sport: Bibliographies and techniques*. Champaign, Ill.: Stipes. **Library call number: Z 6121 Z45.**

Markel, R., Waggoner, S., & Smith, M. (1997). *The Women's Sports Encyclopedia*. New York: H. Holt. **Library call number: GV 709 W665.**

Many university libraries offer virtual as well as real tutorials on Internet research skills and on searching the library's catalogue. You are strongly encouraged to attend a real tutorial, as a librarian in person can anticipate and answer questions outside the box that a virtual tutorial cannot. Finally, most university libraries have an information helpdesk in the reference section, where you can ask for assistance if you get stuck or confused as you do research. You are strongly advised to make use of this facility.

Creative, Original Research

Sometimes assignment instructions state that you should do "creative" or "original" research. What does this mean? It means more than just that you should do the research yourself! It means in addition that you should try to define your topic in terms of subject areas that may not at first seem related. In other words, creative or original research involves showing how the topic has hidden dimensions or multiple perspectives not usually noticed or thought of. For example, our hypothetical topic in section 7.2 above might be regarded as illustrating the beginnings of some creative research. The topic of mandatory physical education at all levels of high school is very often defined in terms of two subject areas: Health, and the bias towards "academic" subjects in high school. You would then research the effects of adequate exercise and lack of exercise on health, and research the status, resources, funding, and the place of physical education in the high school curriculum. You would then likely organize your essay around a thesis something like this one: 'physical education should be mandatory at all levels of high school in order to encourage healthy adult lifestyles'. The connection between health and adequate exercise is well established, and your essay would simply regurgitate that connection with suitable sources to back it up. You would then use the connection between health and adequate exercise as supporting evidence for the above thesis. Noticing that 'theories of education' is another relevant subject area of the topic is the beginning of creative research, because its connection to the topic is seldom noticed.

Keyword Searching Using Boolean Operators

When you use a search engine, whether it is a general Internet search engine or one designed to search a specialized database, you will likely use fields to input keywords or key phrases. At the beginning or end of a field, there is sometimes a pull-down menu giving you the options: "AND," "OR," and "NOT." These are called Boolean operators, with precise logical meanings, and they allow you to create subtle search commands. In order to do effective keyword searches, as well as save yourself a lot of time, you need to understand Boolean operators. Not all search engines present them in the same way – for example, Google presents "NOT" as a field in which you would input keywords to be excluded from the search results. But you need to understand how Boolean operators work in order to appreciate and use the myriad ways in which they are presented by search engines. Below, our hypothetical topic has been used to illustrate their meanings.

AND

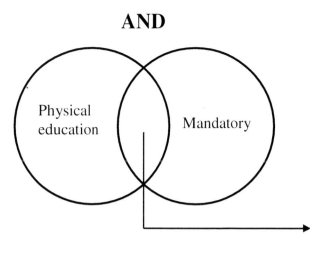

→ Searches for sources that contains both "physical education" and "mandatory."

OR

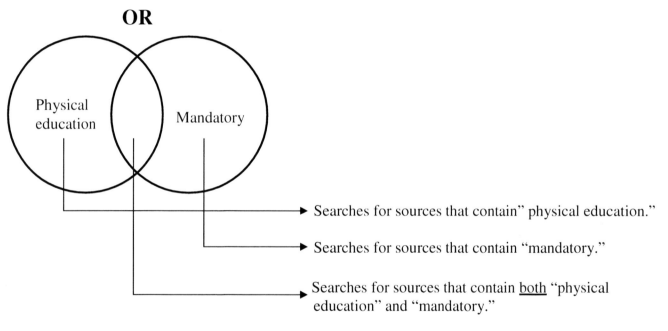

→ Searches for sources that contain "physical education."

→ Searches for sources that contain "mandatory."

→ Searches for sources that contain both "physical education" and "mandatory."

NOT

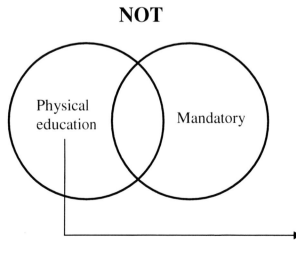

→ Searches for sources that containing "physical education" but not "mandatory."

To see how you could use Boolean operators to create a complex command, consider the following case. Suppose that in addition to searching for sources with the key phrase "physical education" in them, you also wanted to search sources with similar terms for "education," say "training." But imagine that you also wanted to exclude the term "mandatory." How would you do it? With reference to the Boolean meanings above, first, you would select the operator OR for the phrases "physical education" and "physical training." Then you would select the command NOT for the keyword "mandatory" and apply this to the potential results of the OR command. In other words: **("physical education" OR "physical training") NOT "mandatory."** Notice that the NOT applies to the result of the command within the brackets. Most search engines do no require you to think in this cumbersome way about NOT. Usually, there is simply a keyword exclusion field that you would fill in with the keywords or phrases that you wanted to exclude.

The above illustrates how to build a complex keyword search command. It is complex because it contains more than one Boolean operator (it contains OR and NOT). Why is it important to be able to create complex keyword search commands? Part of the art of keyword searching involves identifying and searching for sources with terms and phrases that mean essentially the same thing as the terms or phrases that are given in the formulation of your topic. Every term or phrase has synonyms or near synonyms, and it important to discover them and use them in your keyword searches. For example, key terms in our sample topic have at least these similar-meaning terms:

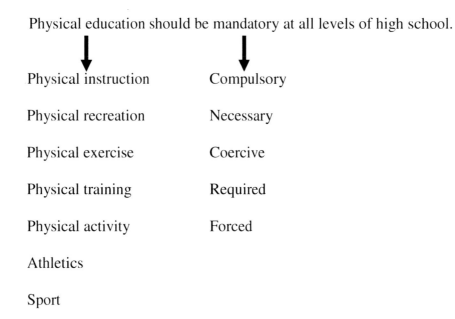

Being able to simultaneously keyword search a whole string of similar-meaning terms while duly excluding sources with unhelpful key terms can cut down on your search time. Most search engines are set up to make the task of creating complex search commands easy. But it is useful to understand the above Boolean principles behind them.

7.4 Locate Suitable Sources

During the course of doing background research, you will likely have used many different types of sources, including popular Internet sites, newspaper articles, magazine articles, journal articles, general interest books, academic books, encyclopedias, dictionaries, bibliographies, indexes, and perhaps even audiovisual sources such as television broadcasts and audio recordings. All these different types of sources will have been invaluable in helping you to define the topic by coming to understand the relevant main subject areas and how they relate to each other. With the topic defined in your mind, you are better equipped to do effective research focused on the types of sources appropriate for academic written assignments. It is possible that while doing background research you might find one or two great sources of a type acceptable for a research paper. Moreover, you may be allowed to base your paper on a wide variety of sources – check the assignment instructions carefully to see if this is the case. But for the most part, you will likely have to discard many sources found during background research as you go forward with a better understanding of the topic to focus more narrowly on high quality academic sources, such as journal articles and academic books.

Why are some types of sources more acceptable than others? A large part of it has to do with the quality of information. Obviously, when writing an academic research paper one wants to base it on information that is not out of date, biased, inaccurate or just plain wrong. One wants to base a research paper on recent, unbiased, accurate data. In other words, one wants the most trustworthy information available. Trustworthiness, however, is a matter of degree, not all or nothing. Some types of sources offer information that can be regarded as more trustworthy than others, partly because different types of sources have different assessment procedures for potential publication of information, and partly because different types of sources have different functions and cater to different kinds of reader.

Academic Journals and Peer-Review

Types of sources regarded as the most trustworthy include academic journals and to a lesser extent, academic books. The most trustworthy academic journals are those with a **double blind peer-review process** for manuscripts offered for publication. This means that the manuscript is assessed by an expert in the subject-area the manuscript is about (this is what "peer" means in "peer-reviewed"), but neither the reviewer nor the author of the manuscript knows who the other is. Hence, a double blind peer-review process is a completely anonymous way of assessing manuscripts offered for publication in a journal. The idea is that the anonymity prevents bias and favouritism from creeping into the review process, ensuring that manuscripts are accepted or rejected solely on the basis of the quality of the information they contain. Blind peer-review, sometimes called **single blind peer-review**, is a similar process, except that the anonymity goes in one direction only – the author of the manuscript does not know who the reviewer is but the reviewer knows who the author is. Double blind and single blind peer-review are generally regarded as ensuring the highest possible quality and excellence of information. Hence, sources that use these procedures are regarded as the best types of sources on which to focus for a research essay or report. It is mainly academic journals that go in for these assessment procedures, which is why they are regarded as the best types of sources to use. Another well-regarded assessment procedure is plain and simple peer-review. Here, the author of the manuscript and the reviewer of it know who the other is. There is no anonymity, but the fact that the manuscript is assessed by a world expert in the subject area helps to ensure that a manuscript is accepted or rejected on the basis of its quality.

How can you tell the type of review procedure a journal uses? Usually, this information is located towards the front or back of an *issue* (an issue is a collection of journal articles published by the journal, usually quarterly – a number of issues making up a year's worth of publications by the journal are often bound into yearly *volumes*). The instructions for potential contributors to a journal are often located outside or inside of the front or back cover of an issue, including information on the type or review process used. Many academic journals have web sites where you can check the type of review process.

Academic Books

Academic books are regarded as very good sources of relatively trustworthy information. An academic book often consists of an author's research, formerly published in an academic journal, organized in a way suitable for a book. Academic books are regarded as sources of relatively trustworthy information because the information is drawn from the author's double blind, single blind or simple peer-reviewed work published in academic journals. However, academic books, by their very nature, tend to be out of date by the time they are published and hit the shelves. They present researched information anything from about six months to several years after the research was originally done and published in academic journals. By the time the book is published, research may have moved on, perhaps further validating, or perhaps modifying or even invalidating the original research that the book is based on. This is why academic books are sources of *relatively* trustworthy information.

Students sometimes ask what the difference is between an academic book and a non-academic book, and how to tell the difference. Part of the answer to this is above: Academic books are often based on research work previously published in peer-reviewed journal articles, while non-academic books are not often based on this kind of thing. Academic books are, in general, aimed at undergraduate and graduate students at university, whereas non-academic books are aimed at general interest readers who may never have been university students. One good way to tell whether a book is academic is to look at the preface and introduction. Often, the purpose of a book and its intended readership are stated here. You will also become familiar with the names of academic publishing houses, such as Oxford University Press, Cambridge University Press, Harvard, MIT, Human Kinetics, and so on.

The pyramid below summarizes the above and provides an approximate indication of where some other commonly encountered types of sources may stand in regard to the trustworthiness of their information.

Plausibility Pyramid

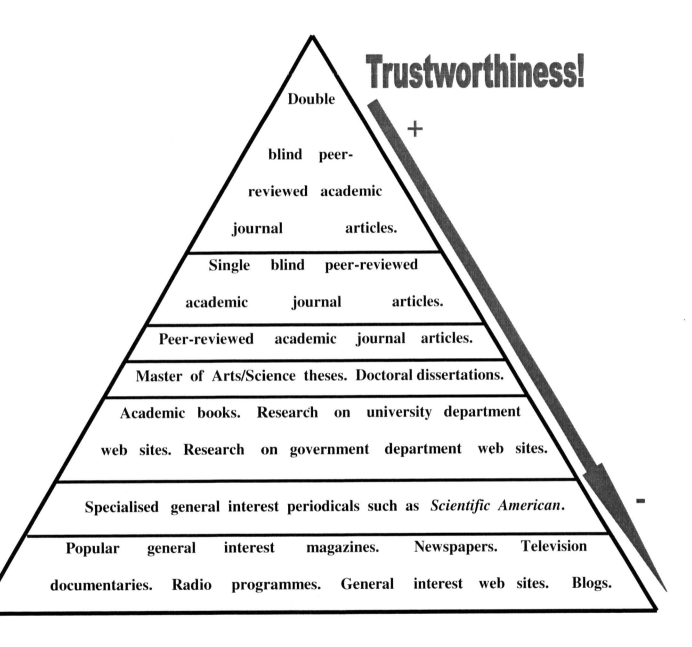

Notice that for your more narrowly focused research, you are really concentrating on sources in the top five layers of the plausibility pyramid. Your assignment may allow you to use sources from the bottom two layers, but if the assignment instructions do not explicitly say that you can, you would do well to check with your instructor. Below are some online electronic indexes and databases that you may find useful as you focus your research on sources from the top five layers. Your university library probably subscribes to some or all of these. Be sure to check.

Web of Science: Emphasizes interdisciplinary science, and it allows you to keyword search scholarly journals from across the academic spectrum.

International Institute for Sport and Human Performance: Allows you to keyword search kinesiology-related master's theses and doctoral dissertations.

Physical Education Index: Allows you to keyword search academic journals and other scholarly sources on a wide range of topics directly and indirectly related to physical education and sport.

Sport Discus: Allows you to keyword search an index of scholarly journals, master's theses and doctoral dissertations, and other types of scholarly sources, on all aspects of sport and recreation.

Medline: Allows you to keyword search scholarly journals on biomedical and health topics.

Elsevier Biobase: Allows you to keyword search scholarly journals on a wide range of biological research topics.

PsychINFO: Allows you to keyword search scholarly sources for a wide range of psychology-related topics.

Sociological Abstracts: Allows you to keyword search scholarly journals and other academic sources on a wide range of sociology-related topics.

7.5 Select Suitable Sources

Research is a little like brain development in that with both processes, one ends up with half of what one started with! If you follow the steps above and thoroughly define the topic, do extensive background research, and then narrow the research to types of sources suitable for the assignment, you will probably end up in possession of a mountain of potential sources of the right types for your paper. In the real world this is not a problem, for there is no cap on the number of sources that may be cited and referenced in a professional researcher's article or scientific report. As an undergraduate student, however, there will very likely be an approximate or exact cap on the number of sources that you are allowed to include in your paper. The reasons for this are many, but one obvious reason is that your written paper has a modest length requirement that must not be exceeded. A professional researcher's article may be as much as six or seven-thousand words long, but the assignment instructions for your essay or report probably stipulate that it must not exceed fifteen-hundred to two-thousand words, or something like that. Moreover, most written undergraduate research assignments require you to discuss <u>all</u> the sources that you choose to include in your paper. In other words, you can't simply discuss one or two in the body of your paper while including an arms-length list of sources on your references page! This means that you have very limited space in which to discuss and analyze the ideas and arguments from your sources. A cap on the number of sources you can use helps to ensure the quality of discussion and prevent lack of space excuses for a cursory or superficial treatment of authors' theses, hypotheses and supporting evidence or arguments. Indeed, it is obvious that the more sources you include in the paper, the less space there will be to treat them all adequately.

How do you narrow down your sources to the number allowed by the assignment instructions? There may be a lot of sources and only limited time in which to select the most appropriate ones. You may have only glanced at the abstracts when you were locating sources, and there may not be time to read them all closely. Hence, the first step is to scan all your sources using the active reading strategies described in section 6 of this manual. When you have completed this step, you are in a position to select sources for your paper on the basis of the following criteria:

- **The most recent sources**. The more recent the source, the more up to date the information is likely to be. The older the source, the more likely it is that the information has been updated or superseded by new findings

- **Select the best sources**. When confronted with a choice of sources, then everything else being equal, choose sources from higher up in the plausibility pyramid. For example, if you have to choose between an academic book and an academic journal article, then if they are similar in all other respects (e.g. equally recent), choose the academic journal article

- **Sources that contain the strongest arguments**. Your written assignment probably requires you to come up with and support a thesis. If you justify your thesis on the basis of critical analysis of weak arguments even though stronger ones are available, you might be vulnerable to the criticism that you have committed a "straw man" fallacy. A straw man fallacy occurs when weaker arguments are chosen for the purpose of being able to criticize them easily

- **Ensure that main perspectives on the issue are represented in your paper**. If the issue you are writing about is a controversial one, don't lose the sense of controversy by choosing sources that all represent only one perspective or argument on the issue! Select sources so that you are able to discuss a range of positions in your paper

- **Sources that contain perspectives or arguments you most agree or disagree with**. It is easier to be engaged with your sources and write a strong paper if you have strongly felt views about the ideas and arguments they contain

You may need to modify or add to the above criteria, depending on the assignment instructions. For example, some assignments require you to include a range of source types, rather than just the best sources. As stated elsewhere, make sure that you read the assignment instructions carefully before making your final selection of sources.

7.6 Evaluate your Selected Sources

Once you have selected material by scanning all your suitable sources using active reading strategies, the next step is to read the selected sources with great care. It is this stage in your research that will enable you to work out or refine an organizational plan for your paper, refine your working thesis if you have one (including abandoning it if necessary), or to identify and develop a thesis. You will probably want to have highlighter pens to highlight important places in the sources, and a notepad and pen with which to make notes. Below are some important points that you need to establish accurately through careful reading of your sources.

- **The author's thesis**. For each of your selected sources, you need to establish with clarity the overall position or claim that the author is trying to establish. Once you have identified the thesis, the rest of the reading will come into greater focus and make more sense. Try to put the author's thesis in your own words in your notebook.

- **The author's main argument or evidence**. For each of your selected sources, you need to identify with precision the main premises or evidence that the author uses to support the thesis. If you do not do this, the critical analysis in your paper will likely be superficial and lack focus by just picking on individual points *ad hoc*. This is a sure sign to the grader that you have not properly understood the source! Try to restate the arguments in your own words in your notebook.

- **Evaluate the author's argument**. For each of your sources, you need to evaluate how well the argument or evidence supports the thesis. To this end, you would firstly want to think about how *trustworthy* the premises or evidence of the argument are – how recent is the evidence, and are there any obvious counter-examples? Secondly, you would need to think about how *relevant* to the thesis are the premises or evidence – is it easy to accept that the evidence is true and yet still see a way in which the thesis could be false? Write down in your notebook the lines of critical thought that lead to your assessment of each argument.

- **Compare the strengths and weakness of the authors' arguments**. When you have evaluated individual arguments as above, step back to compare and contrast their strengths and weaknesses. Write your critical comparison of authors' arguments in your notebook.

- **Think about the 'state' of the issue in light of both how many different main perspectives there are on it and how well these are supported by argument or evidence**. Is there a well-supported dominant view? Can the different perspectives be seen as mutually reinforcing, complementary, or as opposed? Do you see a way in which different perspectives can be accommodated within a broader outlook on the issue?

- **Reflect on where you stand in light of your critical comparison and contrast of authors' arguments**. Who do you agree with, if anyone, and why? This is where you can refine or identify and develop your thesis.

You will notice that much of the really hard, academic work is done during this step, which is the last step in the research process. Students sometimes make the mistake of believing that all the thought and analysis of the issue has to take place at the stage when one is actually starting to write the paper. To be sure, writing a first draft can help further refine your thinking. But **most of the difficult thinking, analysis and reflection is something that you should do, perhaps in note form and with plenty of highlighting of the readings as indicated above,** *as part of the process* **of reading and evaluating your sources**. In other words, most of your hard work will be done by the time you have completed this last stage in the research process, *before* you start to write the paper. Those who try to write their papers before having thought about, analyzed and reflected on the sources often experience writer's block. This is not surprising: How can you know what to write *before* thinking about the sources!

8. THE WRITING PROCESS

There are different kinds of written papers, including essays, reports, position papers, reflection pieces, and case studies. But the same general writing process applies to most of them. **The key idea is that writing is not something you do all in one go.** You don't do your research and then all at once write your completed paper. At least, you don't do this if you are chasing after a passing grade! Writing is a process that involves three main stages: First draft, second draft, and final draft, during which your paper comes progressively into focus. Let's take a look at each of these stages.

8.1 First Draft

Quite often, when students get writer's block it doesn't have much to do with problems concerning the mechanics of writing itself, but to do with incomplete research. "I'm having trouble writing my thesis" is a common concern presented as if it is a writing problem, when in reality it is usually a symptom of not having thoroughly completed all the steps of research. When you have thoroughly completed the research steps explained in section 7, you *will* have some understanding about what your thesis is, and about how your evaluation and comparison of authors' arguments leads to and supports your thesis. Latent in this understanding will be a sense of how information should be organized. The function of a first draft is to try to capture your understanding of all this in the format of your paper (e.g. essay, position paper, or case study) as quickly as possible, while it is still fresh in your mind.

Try to write your first draft briskly, without worrying about spelling, vocabulary, or grammar. These things can be corrected later. You should focus on writing the body of your paper. The introduction and conclusion can be drafted afterwards. Translating your understanding of a topic or issue to written form in a first draft is probably the hardest part of the writing process. Although you should write your first draft quickly, this is not to say that there will be no stops and starts where you have to pause to think about what to say or which way to go. Indeed, writing is itself a mode of thinking, and you will likely find that as you write your first draft you will alter some aspects of your understanding of the topic. Another issue is the sheer fatigue that can be generated. As you write, there will no doubt be a lot of crossing out and re-writing of words, phrases, sentences, and even paragraphs, as you struggle to express yourself. This is simply part of the nature and reality of first drafts, so don't lose confidence in yourself and start to panic when you find yourself frequently crossing out stuff and stalling. Because of all the crossing outs and changes in thinking, writing a first draft is very exhausting, and you should take frequent breaks without feeling guilty. Enjoy the breaks! How long it should take you to complete your first draft depends on how long and involved the written paper is supposed to be. Obviously, you would have some idea about the total time for completing the first draft from having completing your task assessment for the assignment (see section 3.4 of this manual). Depending on the length and involvement of the assigned paper, writing a first draft might take anywhere from a day to a week or more.

A technique that some students find helpful in reducing fatigue in the first draft stage is to create an organizational mind map of the paper before starting to write. It is considerably easier to cross out or change a bubble and key phrase than to cross out and re-write sentences and paragraphs. Playing around with an organizational mind map of the paper can help you think through much of the organization of the paper with relative ease. Let us take an example. Suppose you researched the topic: "Should WADA be scrapped and PEDs allowed in professional sport?" You did all the research and now you are at the first draft stage. Below is an imaginary example of a worked out organizational map for the body of the paper.

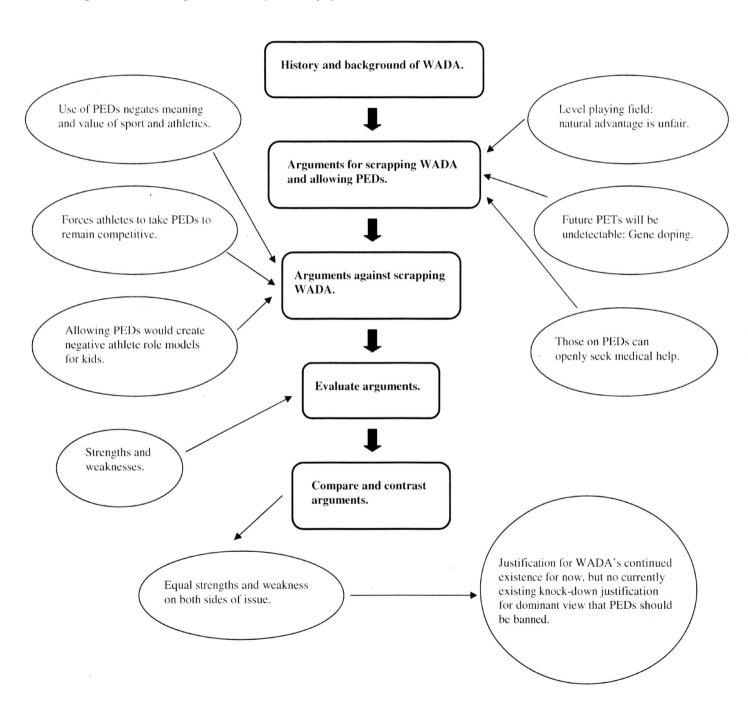

The beauty of using an organizational map before writing is that after you have played around with it and got the organization of information as you want it, the map can function as a helpful guide to follow as you write your first draft.

8.2 Second Draft

Even putting aside such mistakes as spelling, vocabulary, and grammar, it is highly unlikely that your first draft will be free of error. As we write, we believe that what we imagine hearing ourselves saying in our mind's ear coincides with what we have actually written on the page – never was this further from the truth! The second draft stage functions to overcome this illusion by giving you the chance to step back from your writing and assess it.

The simplest way of assessing your first draft is to read it over. There is a danger here, however, because that voice in your head speaking the words as you look at them can continue to create the illusion of a perfect fit between what your written paper actually means and what you think it means. Here are some strategies for overcoming the illusion:

- **Wait for a day or so before reading your first draft**. Things often seem different in hindsight when one is not so close to the event and this is true of written work too. If you put your first draft aside for a day or two before looking at it again, the voice in your head will have less power to create the illusion of a perfect fit when you read it over

- **Read your paper out loud**. Reading your paper out loud allows your brain to process the meaning of sentences and paragraphs using two kinds of stimuli (visual and auditory) instead of just one (visual). This helps to create some distance between you and what you have written, again robbing the narrator in your mind from weaving its spell of a perfect fit between what your written work actually means and what you think it means

- **Get a friend to read your paper to you out loud**. This functions in the same way as the above strategy. An added advantage to having others read out your paper, however, is that they will be reading the sentences cold from the page, whereas when you read it out loud, you already have in your mind assumptions about the emotional tone of sentences, which makes it harder to assess their meaning and impact. Hence, hearing someone else read your paper gives you a less biased, more objective experience of it

- **Get your computer to read your paper to you**. This functions in the same way as the strategy above. But there is a further advantage if you can have your computer read your paper. The addition of human intonation and emotional tone to written work can hide its inherent repetitions and convolutions. The relatively monotone synthesized voice of the computer acts as a magnifying glass that makes you very aware of the rhythms of your sentences and paragraphs, showing up places where you have perhaps been too detailed, not detailed enough, or places where your paper is long and tedious.

What errors should you look for at the second draft stage? You do not need to worry about spelling, vocabulary or grammar, except to the extent that they impact on the following types of errors, which you need to look out for and correct:

- **Wrong meaning.** Sentences or paragraphs that do not mean what you want to say must be corrected or re-written

- **Ambiguous meaning.** Sentences or paragraphs that mean what you want to say but which could be interpreted as meaning something else need to be re-written so that it is hard to interpret them as having a meaning other than the one you intend

- **Redundancy.** Instances of needlessly repeated information, perhaps in the same paragraph or at different places in your paper need to be deleted and the gaps where they occurred closed up, re-writing surrounding sentences to close the gap smoothly if necessary

- **Unclear connection between your thesis and your argument for it.** If it is not clear how your thesis emerges from your analysis of your sources, you need to re-write the body of your paper to make the connection clear. Alternatively, it may be that all you need to do is re-write your thesis-statement in order to make the connection clear

- **Flow.** Does your organization of information ensure that transitions between paragraphs are clear and intelligible? Are there places in your paper where transition from one paragraph to the next might puzzle a reader? If so, you either need to re-write those particular paragraphs or re-organize the information in part of your essay and re-write the entire section. When your paper has good flow, it ought to be possible to move backwards or forwards from any point in the paper smoothly and intelligibly.

A Word of Advice to Perfectionists

The second draft stage in the writing process is a perfectionist's paradise, and it is reported that many students have entered the second draft stage never to be seen or heard of again! There is nothing wrong with wanting your paper to be perfect and desiring to create and demonstrate the best work you are capable of. This is a wholly laudable attitude to have towards your work, and it is to be encouraged. Perfectionism can become pathological, however, and you need to watch out for signs of it. If the second draft stage begins to wildly exceed your estimated completion time for this part of the assignment, if the rest of your life comes to a full-stop, and if things start to lose their point or meaning, you need to stop and recover some perspective. It is normal for professors and graders to spend between fifteen and forty-five minutes maximum reading a two-thousand word paper. The huge amounts of time over the estimate that you take to craft subtle nuances over and above the main arguments and issues are unlikely to be noticed or result in extra points, and may well lead to a lateness penalty, to criticism of having gone off topic, or even to clinical depression. Anybody can be the best given enough time. But the wisest strategy is to just *do your best in the time available* without canceling the rest of your life.

8.3 Final Draft

This is sometimes referred to as the 'polishing stage' in the writing process. Here you need to correct errors of spelling, vocabulary, punctuation, and grammar. You need to check that your paper conforms to the format stipulated in the assignment instructions. It is important to check that in-text citations and reference pages are complete and presented in the correct style. You need to check that direct quotations and paraphrases are correctly presented and duly cited. Finally, you need to ensure that your title page or cover page contains all the relevant information (e.g. your name, student number, course number etc.) and has a meaningful title that does not leave the reader guessing as to the topic and point of your paper. Needless to say, the final draft stage is perhaps the most laborious and boring stage in the writing process. But it is also the easiest stage, and very important, because it deals with the presentation of your paper. First impressions are important. A paper with a silly or meaningless title, egregious spelling mistakes, or which does not conform to format requirements (e.g. it uses an unusual font) creates a negative initial impression on the grader, no matter how brilliant the paper may be in terms of content. Why sabotage all of your hard work by being skimpy or careless at this final, easy stage in the writing process!

<u>Some Common Errors to Watch out for at the Final Draft Stage</u>

Spell-checking and grammar-checking software really helps to cut down time spent at the final draft stage. The software does not catch everything however; indeed, it is surprising how much it does not catch! Therefore, computerized checking should not be the only form of checking that your paper receives. You still need to go through it manually and actively search for mistakes. A good strategy is to use your spell-checking and grammar-checking software at the beginning and end of the final draft stage, with your own, careful manual check in-between.

A handful of Commonly Misspelled Words

<u>Incorrect</u>	<u>Correct</u>
Arguement	Argument
Comitment	Commitment
Consenssus	Consensus
Definate	Definite
Practise	Practice
Procede	Proceed
Seperate	Separate
Supercede	Supersede

Commonly Confused Words and Phrases

Accept
This is a verb.
"I *accept* what you say."

Except
This is a preposition and conjunction.
"The paper is finished *except* for the introduction and conclusion."

Advice
This is a noun.
"Your *advice* is welcome."

Advise
This is a verb.
"I *advise* you to be careful!"

Affect
This is a verb.
"This error will *affect* the results."

Effect
This is a noun.
"The *effect* was startling."

Allowed
This is a verb.
"I *allowed* the test to go ahead."

Aloud
This is an adverb.
"She spoke *aloud* to the people at the back of the hall."

Apart
This is an adverb.
It came *apart* in my hands."

A part
This is a noun phrase.
"This is not *a part* of the plan!"

Because of
This is a verb modifier
"It *melted because of* the heat."

Due to
This is a noun modifier.
They completed it on time *due to* hard *work* and good *time management*."

Could of
Incorrect phrase.

Could have
Correct phrase.

Criteria
This is a plural noun.
These *criteria* are too narrow in scope."

Criterion
This is a singular noun.
The first *criterion* is unclear."

Its
Possessive form of pronoun 'it'.
Its flowers are gorgeous."

It's
Contraction of 'it is' or 'it has'.
"*It's* going to rain!" "*It's* rained!"

Later
This is an adverb.
"She arrived *later* than expected."

Latter
This is an adjective.
"Your *latter* argument is persuasive."

Lose
This is a verb.
"Do not *lose* the equipment."

Loose
This is an adjective.
"The knot is *loose*."

Patients
This is noun.
"Elderly *patients* were mistreated."

Patience
This is a noun.
"Anne's *patience* is her strong point."

Phenomena
This is a plural noun.
"These *phenomena* are puzzling."

Phenomenon
This is a singular noun.
"The *phenomenon* of 'burn out' has been much researched."

Proceed
This is a verb.
"*Proceed* with the plan."

Precede
This is a verb.
"Lightning does not *precede* thunder, although it seems that way."

Should of
Incorrect phrase.

Should have
Correct phrase.

There
This is an adverb.
"The noise came from over *there*."

Their
This is a possessive pronoun.
"*Their* hypothesis was incorrect."

Were
This is a verb.
"The results *were* inaccurate."

Where
This indicates place.
"That is *where* they should be placed."

Whether
This is a conjunction.
"It is hard to say *whether* the results will be decisive."

Weather
This is a noun.
"The *weather* is fine this morning."

Would of
Incorrect phrase.

Would have
Correct phrase.

Your
This is a possessive pronoun.
"*Your* cholesterol level is high."

You're
This is a contraction of 'you are'.
"*You're* lucky to have received such good care."

Common Vocabulary Errors

Problems of vocabulary in students' written work do not usually arise because of a failure to use technical terms. Students develop their technical vocabulary gradually, through exposure to course materials, lectures and tutorials. There is nothing wrong with this as it is a normal part of a student's intellectual development. The types of vocabulary errors we are concerned with have to do with 'styles' of writing.

<u>Informal Words and Phrases</u>
This is the most common form of vocabulary error. Unless your assignment specifically indicates that you may use informal words and phrases, you should assume that you must write in a formal style. Informal written work reflects the way we talk and communicate with family and friends. Words and phrases such as "awesome," "that's sweet," "sort of," "kind of," "I mean like...," and "pretty clear" should be avoided when you write a formal academic paper. When talking with family and friends we frequently use contractions such as "you're," "we're," "she's," "he's," and "I'm." These too are regarded as informal expressions and you should avoid using them in your formal written work.

<u>Jargon Expressions</u>
This is the next most common form of vocabulary error. Statisticians talk of 'percentiles' and 'quotients', geneticists talk of 'clones' and 'recessive traits', and traders talk of 'upturns' and 'downturns' in the market. These are examples of words and phrases belonging to specialized languages. Some words and phrases from specialized languages migrate to everyday language. It is important not to let this happen in your formal written work. For example, you should avoid writing something like: "The ideas of these two authors are *clones* of each other because...," or "the *recessive* point behind the author's assertion is..." or "the *downturn* of this argument is..." Most jargon-words have precise meanings within the specialized languages to which they belong, but lose these precise meanings when they are used as part of everyday talk. In everyday language, jargon-words are rather unclear, and if you use them in your formal written work, your thoughts, ideas and arguments will be unclear too.

<u>Unnecessary Fancy Words and Phrases</u>
This is not quite so common an error, but there are usually one or two papers in every batch with this type of mistake. Graders operate under the assumption that the better a student understands the topic, the easier it will be for the student to express herself in clear, simple, formal language. Graders assume that the less a student understands an issue, the more likely it is that he will resort to using fancy or technical terminology to hide his lack of understanding. Indeed, these assumptions have, by and large proved correct! Resist the temptation to create a tone of professionalism to hide a lack of understanding. This can result in some very ugly sentences, such as: "The psychological predilection of the author for dualistic perspectives on the issue is noteworthy." Just write: "This author is a dualist." This type of writing error creates extra work for graders, who have to spend time deciphering these unnecessarily complex sentences. This ill-disposes graders towards your written work – never a good idea! A far more constructive approach is to turn your lack of understanding to your advantage. Try to be precise about what you do not understand and why, and present your lack of understanding as if it is an analysis of the issue: "The author's theory is unclear at this point because..." Who knows, you might be right!

Common Grammatical Errors

<u>Sentence Fragments</u>

A sentence fragment is a sentence without a main clause. This prevents the sentence from expressing a complete thought. Here are some examples of sentence fragments:

>Traditional male-dominated sports.

>Martin argues that scientific textbooks on human reproduction.

>When we realize that health is a resource.

Can you see how these sentences fail to express complete thoughts? The first sentence introduces the topic of traditional male-dominated sports but fails to go on to say anything about it. The thought is incomplete. The second sentence tells us that Martin argues something about scientific textbooks on human reproduction, but it does not tell us *what* Martin argues. Again, the thought is incomplete. The third sentence tells us that something happens when we realize that health is a resource, but it does not tell us *what* happens! Once again, we have an incomplete thought. Completing these thoughts in additional sentences does not help:

>Traditional male-dominated sports. Help to perpetuate gender-relations in society.

>Martin argues that scientific textbooks on human reproduction. Are full of society's gender-stereotypes.

>When we realize that health is a resource. It is easier to see that the dominant emphasis on traditional health risk factors serves to promote the status quo.

Two sentences with half a thought each does not add up to one complete thought! Indeed, for each of the examples above, we now have two sentence fragments instead of one. The only way to make these sentences express complete thoughts is to join them together:

>Traditional male-dominated sports help to perpetuate gender-relations in society.

>Martin argues that scientific textbooks on human reproduction are full of society's gender-stereotypes.

>When we realize that health is a resource it is easier to see that the dominant emphasis on traditional health risk factors serves to promote the status quo.

Run-On Sentences

Run-on sentences are almost the opposite of sentence fragments. A sentence fragment stops too soon, before the thought has been completed, whereas a run-on sentence doesn't stop soon enough. It goes on, and on, and sometimes on and on and on! In run-on sentences, two or more sentences are joined together without the correct grammatical glue. Here are some examples:

> The medicalization of life has encouraged sports medicine to do more than just focus on rehabilitation and injury prevention sports medicine has been a driving force in the development and promotion of performance-enhancing techniques.

> Coaching education tends to be dualistic because it focuses entirely on training the body but not on equipping the coach with social skills hence traditional coaching does not treat the athlete as fully human but only as a skilled body.

> The *Universal Declaration of Human Rights* was drafted by the United Nations shortly after the Second World War in 1948 the *Covenant on Civil and Political Rights* and the *International Covenant on Economic, Social and Cultural Rights* followed these three human rights documents laid the groundwork for subsequent human rights thinking and practice.

To detect and correct run-on sentences, you need to look closely at any sentence you write that seems to express more than one complete thought. Make sure that appropriate conjunctions and punctuation are present to correctly glue two completed thoughts together, or break the complete thoughts up into more than one sentence. The examples above are corrected below:

> The medicalization of life has encouraged sports medicine to do more than just focus on rehabilitation and injury prevention. Sports medicine has been a driving force in the development and promotion of performance-enhancing techniques.

> Coaching education tends to be dualistic because it focuses entirely on training the body but not on equipping the coach with social skills. Hence, traditional coaching does not treat the athlete as fully human but only as a skilled body.

> The *Universal Declaration of Human Rights* was drafted by the United Nations shortly after the Second World War, in 1948. The *Covenant on Civil and Political Rights* and the *International Covenant on Economic, Social and Cultural Rights* followed. These three human rights documents laid the groundwork for subsequent human rights thinking and practice.

Notice that the corrected examples are clearer than their run-on counterparts. As a general rule, wherever possible, always break longer sentences up into shorter ones. Short sentences tend to be easier to read and understand than longer sentences.

The Passive Tense

There is nothing wrong with using the passive tense. Not mentioning the subject of an action can create a useful tone of objectivity, as, for example, when describing the materials and methods of an experiment. Here, using the passive tense can help focus the reader on the actions or procedures followed rather than on the people performing or following them. Unnecessary overuse of the passive tense, however, can detract from the clarity and power of your writing, because it takes more words to express an idea in the passive tense than in the active tense.

In the active tense, the subject of the sentence performs the action of the verb. In the passive tense, the subject of the sentence receives the action of the verb, or the performer of the action is not mentioned at all. Here are some examples to help make this distinction clearer:

Active Tense

Descartes argues for the dualism of mind and body.

This essay demonstrates that within normal limits, testosterone cannot be regarded as causing aggressive behaviour.

Passive Tense

The dualism of mind and body is argued for by Descartes.

In this essay, it will be demonstrated that within normal levels, testosterone cannot be regarded as causing aggressive behaviour.

Always write the introduction of your paper in the active tense, as the passive tense creates a sense that the paper is uncertain and timid about what it is trying to establish. You can use the passive tense judiciously at places in your writing where you need to describe actions without distracting the reader's attention with details of who or what performs the actions.

Number Disagreement

A grammatical error occurs when a verb fails to agree in number with its subject, or when a pronoun fails to agree in number which what it refers to. This is a common error that can cause confusion and ambiguity. It is also distracting for the reader. Here are some examples of this type of error:

Their **arguments** for attracting more medical students to the field of gerontology **is** very sound.

These **nursing practices** need to be reviewed. Indeed, **it** needs careful examination from the perspective of the Charter.

In the first example above, failure of number agreement between subject and verb creates ambiguity about whether the sentence means one or several arguments for attracting medical students. In the second example, failure of number agreement between a pronoun and what it refers to creates ambiguity about whether the sentence means one or several nursing practices. Watch out for this type of error. Here is how our examples look when we correct them:

> Their **argument** for attracting more medical students to the field of gerontology **is** very sound.

> These **nursing practices** need to be reviewed. Indeed, **they need** careful examination from the perspective of the Charter.

Tense Ambiguity

Tense indicates whether an action takes places in the past, present, or future. When tense is muddled in a sentence, ambiguity occurs. Here is a simple example: "John **will go** to the match and then he **went** home." Does this sentence mean that John decided to go to the match, and that after deciding this is what he would do (and without actually having done it yet) he went home? Does the sentence mean that John will go home after having gone to the match? Does the sentence mean that John actually went to the match and then went home? It is hard to tell because of tense disagreement between "will go" (future) and "went" (past). Making these verbs agree in tense removes the confusion: "John **will go** to the match, and then he **will go** home," or "John **went** to the match, and then he **went** home." Tense ambiguity can occur across sentences, not just within single sentences. Here is an example:

> This paper **will argue** that the Olympic Games has always failed to live up to the ideals of Olympism. It **has been demonstrated** that the Games are doomed to become just another commercial athletic event.

The first sentence sets the action of arguing in the future. But the second sentence sets the very same act, or part of it, in the past – "demonstrated" refers to the same act of arguing referred to in the first sentence. This tense disagreement creates the impression of imprecision and sloppy thinking. As a matter of convention, it is appropriate to keep the actions of a paper in the present:

> This paper **argues** that the Olympic Games has always failed to live up to the ideals of Olympism. It **is demonstrated** that the Games are doomed to become just another commercial athletic event.

Other Common Errors

Faulty Presentation of Direct Quotations and Paraphrases

Direct Quotations

Kinesiology uses *American Psychological Association* (APA) conventions as its style guide, and APA stipulates how you are to present direct quotations and paraphrases. A direct quotation repeats what an author has written in exactly the same way as the author has written it. **"Direct quotations of *less than forty words* must be included among the sentences of your paragraphs but must be enclosed by double quotation-marks, just as this sentence is"** (Ashby, 2007, p.54).

Direct quotations of *forty words or more* must be indented and blocked, without quotation-marks. Indented block quotations must be double-spaced, not single-spaced. When you read books and articles, you will come across block quotations that are single-spaced. Single-spacing is acceptable for published works. But your paper is not a published paper! So you must double-space all your block quotations. Here is an example:

> **Direct quotations of *forty words or more* must be indented and blocked, without quotation-marks. Indented block quotations must be double-spaced, not single-spaced**. When you read books and articles, you will come across block quotations that are single-spaced. Single-spacing is acceptable for published works. But your paper is not a published paper! So you must double-space all your block quotations, just as this one is. (Ashby, 2007, p. 54)

You must also cite the source <u>whenever</u> you present a direct quotation. For more on citing, see the section on APA citations and references in this manual.

Direct quotations are useful when an author's exact wording of a particular idea or issue is crucial to your argument or analysis. But do not use direct quotations simply to avoid explaining in your own words ideas or issues that you find difficult – this would lose you marks. Graders are impressed when they see you struggling to explain ideas yourself rather than lazily resorting to direct quotes.

Paraphrase

You paraphrase when you put what an author has written into your own words and sentence-structures. **You should aim to put an author's ideas *entirely* into your own words and sentence-structures.** This is called a loose paraphrase. When you use loose paraphrase to express an author's idea, you should not enclose it in quotation-marks, and you should include it among the sentences of your paragraphs. There is also no such thing as a block paraphrase, so you should not indent paraphrases of forty words or more long! **But you must provide a citation for where the idea you are paraphrasing comes from.**

Close paraphrase, where one expresses an idea using many of the author's own words and sentence-structures, **is not acceptable**. At university, close paraphrasing is regarded as a form of plagiarism (cheating). When you give citations for sentences not enclosed in double quotation-marks, you are essentially signaling to the reader that the wording and structure of the sentences are entirely your own (loose paraphrase). Hence, on occasions where you find you can't come up with a loose paraphrase of an author's idea, you must give a direct quotation, not a close paraphrase.

Faulty Reference Lists

A reference list gives full details of the sources you have cited in your paper. If your paper is for a kinesiology course, you must use APA style. The references list is not an unimportant part of the paper that you tack on at the end quickly. The references list helps the reader to see how you have contextualized your work, in addition to allowing the reader to locate and read sources you have used. Graders take references lists (and citations) very seriously, and marks can be lost for faulty style or incompleteness. Hence, you should scrutinize your citations and references for errors and correct them before printing out and handing in the final draft of your paper. For more on references lists, see the section on APA citations and references in this manual.

Format Errors

Format concerns the general style in which your written work is presented. This general style covers such things as the information you must include on the title or cover page, if and how to number the pages of your work, font size and type, margin widths, and line-spacing. The format instructions will likely be in the course outline you receive at the beginning of a course, or they may be written into the assignment instructions. **Follow them exactly**. If they do not appear anywhere, you should enquire. After all your hard work on the paper, format seems like such a trivial issue. But you are well-advised to take it seriously. Correct formatting of your written work adds polish to its presentation. Also, marks can be deducted for format errors. Why sabotage your efforts at the last hurdle? Go all out for the best grade!

9. ESSAYS

When students think of written assignments, they immediately think of essays. To be sure, the essay is the most assigned form of written work. There are many different types of essays, but most of them share basic structural components and features. Here, we will look at four of these basic components and features: The introduction, the body, the conclusion, and the thesis-statement.

9.1 The Introduction

The *introduction* comes at the beginning of the essay. The purpose of the introduction is to prepare the reader for the body of writing that comes after it. Having completed all the steps of research, you know what you are writing about, and what your thesis is. But unless you inform your readers of this in an introduction, they will feel lost and judge your essay to be an unclear, disorganized piece of work. A good introduction includes three elements:

- Indicates clearly a well-defined topic for the essay

- Describes briefly how the body of the essay is organized

- Explains the point of writing the essay.

The point of writing an essay is usually to argue for a thesis. So you need to briefly indicate your thesis and how you argue for it. This is called a thesis-statement, and most essay introductions include one.

Let's illustrate the above with an example. Suppose you had to write a ten-page essay on the topic of whether body checking should be completely banned in minor league ice-hockey. You did your research and found that there are several main arguments for and against a total ban. In the body of your essay you described and evaluated these arguments, and determined that arguments for a total ban are stronger than arguments against a total ban. Now you must write your essay's introduction! Here is how a student new to academic essays might write the introduction – for illustrative purposes, the topic sentence is in bold, sentences on the organization of the body of the essay are in italics, and the thesis-statement is in regular font:

> **This essay is about the issue of body checking in minor league ice-hockey.** *First, arguments for a total ban on body checking are examined. Second, arguments against a total ban are discussed.* It is shown that pro-ban arguments are stronger than anti-ban arguments. Therefore, the thesis of this essay is that body checking in minor league ice-hockey, even on a limited experimental basis, should be banned.

This introduction is all right so far as it goes. It is better to have an introduction that includes the three important elements than one that does not. This introduction also has the virtue of being clear, and clarity is essential. But let us review it to see if it can be improved. The topic sentence could be improved. Instead of writing: "This essay is about..." it would be better to write a few sentences that convey a sense of the current state of the topic. This not only tells the reader what the topic of your essay is, but it also gives the impression that you are knowledgeable about the topic and in command of your research material. Next, notice that the organization sentences use the organizational words "first" and "second" to help describe how the body of the essay is structured. However, if you can convey the structure of your essay without using too many organizational words, that is even better. Finally, the thesis-statement could be improved. Rather than writing: "Therefore, the thesis of this essay is..." simply give a bold, factual sentence that expresses your position on the issue. This conveys an air of confidence, unlike "...the thesis of this essay..." which comes across as timid and non-committal. Following the same stylistics as above for illustrative purposes, here is how the above introduction might read after correcting these errors:

> **Body checking has always been a controversial issue. However, the decision of Hockey Canada to allow some hockey associations to permit body checking on an experimental basis among players as young as nine years of age has aggravated the controversy quite considerably.** *Perspectives on the issue fall into three main categories: Viewpoints of fans, the official standpoint of Hockey Canada, and positions held by the scientific community.* Evaluation of the main arguments shows quite clearly that the decision to allow body checking in some minor league games, even on a limited experimental basis, is a serious mistake.

In this improved introduction, notice that the topic sentences give an impression of the current state of the topic, and so convey the topic of the essay to the reader, without using the words "essay" or "topic." The organizational sentences convey the structure of the body of the essay using few obvious organizational words. The thesis-statement shows clearly where you stand on the issue and how you arrived at your position, without using the words "essay" or "thesis."

The improved introduction gives the reader the impression that you are knowledgeable on the topic, and that doing the research led you to an intelligent, informed decision. Why doesn't the first introduction example have the same effect? Words like "essay" and "thesis" make it seem as if there is a gap between you, the writer, and the essay itself. This gives the impression that the concerns about and position on the issue may not be *your* concerns and position – just the essay's! Notice that the improved introduction gives the impression that there is no gap, and that you are expressing yourself *through* the essay.

Practice writing essay introductions without using phrases such as "the topic of this essay" or "the thesis is..." Also try to write topic sentences without using phrases such as "the topic of this essay is...," "this essay focuses on...," or "this paper is about..." This may be particularly challenging because it is easy to include too much detail and end up with an unintended body paragraph instead of an introduction! But with practice, you will be able to write more effective introductions.

Some Frequently Asked Questions

How long should my introduction be? A common mistake is to write an introduction that is too long – the introduction is so detailed that it is indistinguishable from the body of the essay! As a very rough guide, and there are exceptions, try to make sure that your introduction does not exceed approximately 8% of the total length of the essay. For example, the introduction of a ten, fifteen, and twenty-page essay would be roughly a page, a page and a quarter, and a page and a half respectively.

How detailed should my introduction be? Sometimes the introduction of an essay can be so detailed that it fails to convey clearly the topic of the essay. The introduction only needs to state the topic, general organization, and the thesis of the essay. The longer the essay is supposed to be, the more detailed the topic, organization, and thesis sentences can be.

Why am I finding it hard to write my introduction? The introduction must indicate the topic, organization, and thesis of the essay. If you are not clear about any one of these elements, you will find it hard or even impossible to write your introduction. Writer's block can happen when you try to write your essay without having thoroughly completed all the steps of research. How can you know what your thesis is until you have done all the research! How can you know what the organizational structure of the essay will be until you have completed your first draft of the body! *To save time and avoid writer's block, always write the introduction last.*

Is an introduction like a summary? An introduction is not a summary. A summary repeats the main ideas of an essay. An introduction does not do this. Instead, it indicates the topic, organization, and thesis of the essay.

How many paragraphs should I use for my introduction? The introduction needs to indicate the topic, organization, and thesis of the essay. In a short, five-page essay it should be easy to include all of these elements in a single paragraph. In a longer essay, you can be more detailed about these elements and may need to give them separate paragraphs. An introduction does not have to be a single paragraph.

9.2 The Body

The body of the essay is where all of your hard work during research is reflected, as you contextualize the topic or issue, discuss and evaluate main lines of thinking, and establish your thesis. First, you need to contextualize the topic. How did the issue arise, and when? Has thought on the issue shifted over time? What other main issues does this one directly connect with? Next, you need to discuss and evaluate current main lines of thinking on the issue. Describe the main current perspectives on the issue and the arguments or evidence supporting these perspectives. Note any stark similarities or differences among the perspectives and/or the arguments and evidence supporting them. Once you have described the main perspectives and supporting arguments or evidence, you need to evaluate the perspectives by evaluating their supporting

Body of Essay

Contextualize the topic or issue
What is the history of this issue? How has thought on the issue shifted over time? What other main issues does this one directly connect with?

Discuss current main perspectives
Describe the current main perspectives on the issue and their supporting arguments and evidence. Note any strong similarities or differences among the perspectives or supporting arguments/evidence.

Evaluate current main perspectives
Evaluate arguments and evidence for the main perspectives on the issue. Note which perspectives, if any, are better supported.

Your thesis!
Your thesis should emerge from and be supported by the findings of your evaluation of the main perspectives on the issue.

---

arguments or evidence. You may find that one perspective on the issue is better supported by evidence than the others. You may find that they are all equally well supported. Or you may find that none of them are well supported! Whatever the results of your evaluation, the important thing is that the body of your essay should give the impression that your thesis emerges naturally as a result of your evaluation. You should avoid giving the impression that you evaluate different perspectives on the issue through the perspective of your own thesis, as this can make it seem as if your evaluation is biased and prejudiced.

Organizational Patterns

With so many perspectives and pieces of information to put in place, organizing the material can seem like a daunting task, even though you likely have some sense of how things should be arranged. As mentioned earlier in the section on the writing process, creating a mind map of the body of your essay and then using it as a writing guide can be a relatively painless way of working out its organizational structure. However, it may be difficult to create a mind map if you don't know which types of organizational patterns to impose on different kinds of key points. There are three types of organizational patterns: Spatial, chronological, and logical. Think about how each of these types of patterns could be useful for organizing different kinds of information gathered during the course of researching the topic of whether body checking should be totally banned in minor league ice-hockey.

Spatial Patterns

If your essay involves description or evaluation of a phenomenon or event that is spread out in space, you can impose a spatial organizational pattern on your key descriptive or evaluative points. For example, you might use the spatial structure of the hockey rink, with its specific designated areas, as a way of organizing key descriptive points in your description of how body checking occurs. In other words, you would impose the spatial structure of the rink onto your key points in order to give a clear description of how body checking occurs on the ice.

Chronological Patterns

If you are writing about something that is made up of events or stages, you can impose a chronological pattern on the key points. Simply use terms like "first," "next," "afterwards," "before" and "then" to impose an 'earlier–later' pattern on the key points leading up to or constituting the event. Take the body checking example again. Think about the events that make up an occurrence of body checking. What, typically, happens first? Then what? How does it end? Take another example. Suppose that later in your essay you went on to discuss some of the neurological processes involved in Second Impact Syndrome. Again, you could impose a chronological earlier–later pattern on the key points of the neurological processes involved. When you organize key points by organizing them according to stages that unfold in time, you are imposing chronological organization on them.

Logical Patterns

Logical patterns are patterns of meaningful connection between ideas. As you organize key points in the body of your essay, some key points will have closer meaningful connections to each other than other key points, and some key points will more naturally follow other key points conceptually. Organizing key points along these lines is to impose a logical pattern on them. Failing to organize and sequence conceptually related key points together can lead to very unclear discussion and evaluation. Hence, once you have imposed relevant spatial and chronological patterns on your key points, you should further organize your material according to a logical pattern. Let's use our body checking topic to illustrate how to impose a logical pattern on your key points. Suppose that you had done your research on the topic of whether body checking in minor league ice-hockey should be completely banned. Below are the key points you discovered through research.

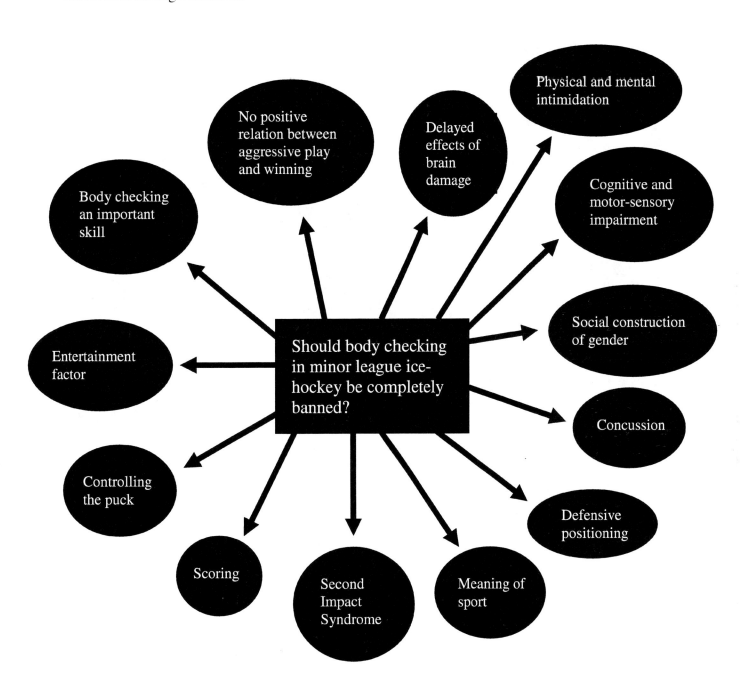

A logical pattern has not yet been imposed on the key points above. First, points that are particularly relevant to each other must be grouped together to form sub-issues. Second, the points within each sub-issue must be organized into a sequence that makes most sense. The result of performing these two steps is below.

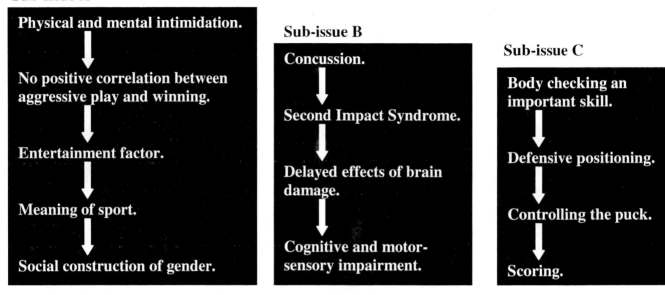

Finally, the sub-issues must themselves be sequenced so that transitions from one to the next are as clear and intelligible as possible. The sequence: sub-issue C, followed by sub-issue B, followed by sub-issue A seems to offer the clearest transitions. Sub-issue C includes explaining what body checking actually is and the officially given reasons for allowing it. This needs to come first in order to provide a context for sub-issues B and A. How can the reader know why sub-issues B and A are relevant to the question of body checking if the reader has not yet been told what body checking actually is and why people want to allow it in minor league ice-hockey? Sub-issue B discusses the possible serious negative medical effects of body-checking on the developing brains of young players. Hence, sub-issue B problematizes the practice of body checking. Sub-issue A discusses the nature and meaning of sport from the perspective of problematized body checking, which latter is therefore assumed as an already established context. For this reason, sub-issue B needs to come before sub-issue A.

Combined Organizational Patterns

Usually, the body of an essay is best organized using a combination of organizational patterns. For example, thinking about the body checking topic, you would use spatial patterns to organize key descriptive points concerning how body checking occurs, chronological patterns to describe neurological processes involved in concussion and Second Impact Syndrome, and logical patterns to organize transitions between sub-issues.

9.3 The Conclusion

Students sometimes neglect to put much effort into the conclusion on the mistaken assumption that it isn't that important compared to the rest of the essay. In fact, the conclusion performs several crucial functions:

- **Brings a sense of closure to your work**. A well-researched and nicely written essay that suddenly stops without warning takes away from the sense of organization and direction that you have worked so hard to achieve in the body of the essay

- **Summarizes the essay and tests your understanding**. The conclusion is essentially a summary of what you take to be the main points and issues, and of how you think they lead to and support your thesis. It is hard to summarize a piece of writing that you do not fully understand. For this reason, graders are particularly interested to see if the conclusion has been done well

- **Suggests further lines of research in light of your thesis**. What further research does your thesis imply? What questions does it raise? Addressing these questions briefly in your conclusion demonstrates that you have thought deeply about the issue. However, be careful not to introduce new main ideas or concepts that you have not discussed in the body of your essay.

Below is an example of a conclusion that might have been written for our 'ban on body checking' topic:

> The question of a total ban on body checking in minor league ice-hockey will remain a contentious issue for the foreseeable future. The possible negative, life-long cognitive and motor-sensory impairment of repeated concussion is well-established. Aggressive, forceful play is not just about controlling the puck, however, but it connects with deeper issues in society concerning the construction of gender and empowerment. **[Summary of main points.]** This is why there is resistance to acting on the clearly demonstrated medical risk factors. Until dominant constructions of masculinity and femininity are successfully challenged and transformed, there will be continued, intermittent pressure to incorporate medically dangerous aggressive play, including body checking, into traditionally male team sports at the junior level. **[Your thesis.]** This raises questions about the possible need for wider powers of government to monitor and where necessary, to veto initiatives shown to be medically harmful. A policy taskforce should be set up to explore this option. **[Further research.]**

Notice how the conclusion strikes a note of closure, and summarizes main points concisely, without repeating all the details discussed in the body of the essay. Being able to summarize in this way requires practice. Notice also how a further line of enquiry is offered without introducing new ideas that would affect or upset the already laid down focus of the essay.

9.4 Theses and Thesis-Statements

What is a Thesis?

Most essay assignments require you to develop and defend a thesis. Good grades typically depend on well-chosen, well-argued theses. But what is a thesis exactly? Technically, any claim about something is a potential thesis. But for the purposes of an academic assignment, a thesis is usually a controversial claim or viewpoint, which means that it needs to be defended by argument or with evidence.

There are two main sorts of theses: working theses, usually called working hypotheses, and theses that emerge as a result of research, which we will call 'emergent theses'. When one has a successful working hypothesis, the two types of thesis coincide. A working hypothesis is a hunch, or a temporary position that one adopts in order to give one's research an initial focus and direction. In other words, it is a way of 'kick-starting' one's background research. A working hypothesis may be supported by the research and remain as one's final, emergent thesis, but it is also possible that the working hypothesis will need to be altered or even dropped in favour of a different thesis if the research does not support it. An emergent thesis is a main position or viewpoint that one comes to adopt after having done the research and evaluated the research sources. Here, there is a naturally close conceptual connection between the thesis and the findings of one's evaluation of the research sources. One can see how the thesis emerges from the findings. Students sometimes make the mistake of adopting a working hypothesis and then selectively locating sources to support it without doing unbiased, extensive background research. This results in an artificially close connection between what remains essentially a working hypothesis and some selected sources. This affects an essay's grade. For the working hypothesis to become an emergent thesis, one would need to research widely and without bias, and then discover that one's working hypothesis is supported by the results of evaluating an unbiased selection of research sources.

How Good is My Thesis?

There are four main ways in which one's thesis can be flawed. All four types of error are quite common. We will review them here, so that you can avoid these mistakes.

No Connection
Suppose you have an assignment that requires you to write an essay on whether mandatory physical education at all grades of high school would significantly impact on adult obesity levels in Canada. You come up with the thesis: Eating disorders among high school students contribute significantly to high obesity levels in adult life. The thesis is interesting but notice that it has nothing to do with the assigned topic, which concerned mandatory physical education at all high school grades, not eating disorders at high school! **The thesis bears no connection to the assigned issue. This is sometimes called** *going off topic.* Another way of going off topic is to discuss evidence that has no connection to the thesis you indicate in your thesis-statement at the beginning of the essay. If your stated thesis is that there should be mandatory physical education at all high school grades, don't spend your essay discussing evidence for the thesis that eating disorders at high school contribute to adult obesity!

Trivial Theses

Suppose you have an assignment that requires you to write an essay on violence in sport. You come up with the thesis: Violence is indeed a problem in sport. This would be an example of a trivial thesis. **A trivial thesis is a claim or position that almost everyone already agrees with, making the task of supporting such a thesis in an essay pointless.** Almost everyone agrees that violence in sport is a problem, so arguing or this claim is a waste of time because you don't have to convince anyone. Notice that this thesis is also trivial in regard to the assignment instructions. If you are asked to write on the issue of violence in sport, then it is fair to say that it is a given that violence in sport is a problem! The opposite of a trivial thesis is a controversial thesis. Controversial theses are much harder to argue for. This is why essays containing controversial theses usually receive higher grades than those containing trivial theses. To avoid trivial theses, make sure that your thesis is more detailed and specific than any assumptions about the topic that are built into the assignment instructions.

Wide Theses

A thesis is too wide if it claims more than is necessary. For example, suppose that you are required to write an essay on access to public recreational facilities in Toronto. You come up with the thesis: The root causes of lack of access to public recreational facilities are the same in the Atlantic and Central Canadian provinces. This thesis does bear a connection to the topic of the essay, but its scope includes more than was assigned (access in Toronto). Wide theses should be avoided because they leave your essay open to the criticism of having gone off topic, and because they are much harder to defend – why create extra work for yourself!

Narrow Theses

A thesis is too narrow if it claims less than is necessary. Suppose you are required to write an essay on the changing nature of the Olympics over the last one-hundred years. You come up with the thesis: The commercialization of the Olympics is a recent phenomenon with beginnings traceable to the 1984 Los Angeles "Hamburger Games." This is an interesting thesis to be sure. And it does bear a connection to the assigned topic. But it is too narrow. Its scope includes less than was assigned (changes in the Olympics over the last <u>one-hundred years</u>, not just over the last twenty-five years!). Avoid narrow theses because they leave your essay vulnerable to the criticism of having gone off topic, and of having failed to address the topic in sufficient depth.

<u>How to Avoid Poor Theses</u>

To help avoid coming up with a thesis that is irrelevant, trivial, wide, or narrow, make sure that you fully understand the assignment instructions and allow yourself plenty of time to research the topic. As you compose your first and second drafts, continually check that the evidence you discuss actually supports your stated thesis rather than supporting a subtly different thesis that develops in your mind as you write. If necessary, reformulate your stated thesis in order to reestablish or strengthen its connection with the supporting evidence. Ask yourself how the thesis addresses the assigned topic in its entirety.

What is a Thesis-Statement?

At its simplest, a thesis-statement is just a sentence, or a couple of sentences that tell the reader what the thesis of your essay is. Thesis-statements are factual sentences like any other factual sentence. A thesis-statement typically appears in the essay's introduction, in the body of the essay as it emerges from your discussion and evaluation, and in the conclusion. Do not confuse a thesis-statement with a thesis. A thesis-statement is as easy to write as any other factual sentence. The hard part of a thesis-based assignment is discovering a good thesis and defending it effectively with strong reasons or evidence.

Some Frequently Asked Questions

Why am I finding it hard to come up with a thesis and thesis-statement? The most common reason for this is that you are trying to think of a thesis before having done sufficient reading and research. It is difficult to come up with an effective thesis about an issue until you have thoroughly researched that issue! Do the research first and it will then be easier to discover a good thesis. Once you have hit upon a thesis, it will be quite easy to write a thesis-statement.

How long should my thesis-statement be? Thesis-statements vary in length, depending on the complexity of the issue and the nature of the assignment. But in a typical undergraduate essay assignment, the thesis-statement is not usually expected to be more than a sentence or two long. Your thesis-statement needs to be clear for the reader, so keep it straightforward – avoid complicating details when stating your thesis in the essay's introduction. Save the details for the body of the essay.

Supporting Your Thesis

To support your thesis you must argue for it, and that means providing evidence to persuade your reader of its truth. How well you support your thesis is an indicator of how well you understand the topic of your essay. You would not get the best grade for an essay with a badly supported thesis, no matter how interesting the thesis itself was. Ideally, your thesis should be seen to be supported by the results of the discussion and evaluation in the body of the essay. Overall, if your supporting argument is to be effective, it must include *trustworthy* evidence, and the evidence must be *relevant* to the thesis.

Trustworthy Evidence

Trustworthy evidence is usually recent. Old data has likely been replaced by newer information. Secondly, trustworthy evidence comes from reputable sources. A double blind peer-reviewed journal article is more reputable than an academic book, and an academic book is more reputable than a popular Internet site, magazine or newspaper (see section 7.4 and the Pyramid of Plausibility for more details). Finally, trustworthy evidence is plausible evidence – it should be hard to criticize.

Relevant Evidence

If the evidence is to lend credibility to your thesis then it must be relevant to the thesis. Relevance, like trustworthiness is a matter of degree. There is a golden rule that you should use when assessing how relevant a piece of evidence is to your thesis:

Golden Rule

The easier it is to imagine that the evidence is true but that the thesis is false, the less relevant the evidence is to the thesis.

Using this rule takes some practice. But it is worth it, since it is at the very heart of what it means to "evaluate." Finally, note that it is possible for evidence to be relevant to your thesis but not trustworthy, or to be trustworthy but not relevant to your thesis. Be careful to ensure that the evidence you use to support your thesis is *both* trustworthy *and* relevant!

10. SOME OTHER KINDS OF ACADEMIC WRITING

Apart from the essay, there are several other sorts of written academic assignment. Some of the more common sorts are briefly reviewed in this section.

10.1 Writing an Abstract

An abstract is a short, complete summary of a piece of written work. The purpose of an abstract is to save time. It helps the potential reader of the work decide whether it is worthwhile investing time reading the entire article, report, or book. Most academic journal articles have abstracts at the beginning, and it is worth studying some of these to gain an impression of what a professional abstract is like.

Characteristics of a Good Abstract

A good abstract has the following features:

- **Self-contained**. The abstract should not require the reader to look beyond it in order to understand the summary it contains

- **Short summary of the entire work**. It should summarize the topic, method used to establish or support the thesis, the results or thesis, and the main implications. An abstract may be just a few sentences or up to 250 words or more long. If you are given an abstract assignment, pay close attention to the length requirement

- **Clear and concise**. It should contain no ambiguity or unnecessarily wordy sentences

- **Objective**. The abstract should not express the personal thoughts, feelings, or beliefs of its author

- **Fixed scope**. It should not include any information not already included in the entire work. It should not introduce anything that is not in the entire work

- **No odd abbreviations etc**. The abstract should contain no uncommon abbreviations or acronyms, citations, figures, or tables. It should contain no element that requires the reader to look beyond the abstract in order to understand that element.

The standard content of an abstract is given by the second bulleted item above. However, the standard content is sometimes modified for the purposes of an assignment. Read the assignment instructions carefully.

Keeping your Abstract Short

The top two difficulties with writing abstracts are 1) letting go of details to present only main ideas, and 2) using as few words as possible to express the main ideas. There are four common errors to watch out for while trying to stay within the length requirement of an abstract assignment:

- Do not repeat in different words what you have already stated
- Do not explicitly state what is obviously implied by a concept or idea you have already introduced
- Do not explicitly state what is obvious from the already established context
- Do not assume that separate sentences are necessary to summarize different sections of the work. Often, summaries of two sections of the work can be worked into a single sentence.

Look over the following attempt at a 50 words limit abstract, which has failed to avoid the four common mistakes just mentioned, and consequently exceeded the length requirement by 67 words! Then look at the corrected version after it. Think about why the changes were made.

This paper is about soccer violence by fans. It focuses on the cause or causes of soccer fan violence before, during and after the game. A prevalent theory is that committing violence is a strong motivating factor to attend the game and hence, that soccer violence among fans is pre-planned. However, recent research suggests that pre-planned violence is not the cause of soccer hooliganism committed by fans. The evidence for these competing views shows that the recent research is more persuasive than the pre-planned violence theory. This has implications for how soccer violence among fans is addressed. In particular, the rejection of the pre-planned violence theory has implications, both for crowd control strategies and for stadium design.

This paper argues that ~~causes of soccer violence by fans, and after the game. A prevalent theory is that committing violence is a strong motivating~~ factor to attend the game and hence, that soccer violence among fans is pre-planned. ~~However,~~ recent research convincingly shows that pre-planned violence is not the cause of soccer hooliganism committed by fans. ~~The evidence for these competing views shows that the recent research is~~ more persuasive than the pre-planned violence theory. This has implications, ~~for how soccer violence among fans is addressed. In particular, the rejection of the pre-planned violence theory has implications,~~ both for crowd control strategies and for stadium design.

10.2 Writing a Synopsis

A synopsis is a more detailed summary than an abstract. It adds meat to an abstract's skeletal description by providing more details of the main ideas, supporting evidence, implications, and how they transition and connect with each other. In other words, a synopsis gives a good impression of the 'character' of a piece of work. Some synopses provide a purely descriptive summary, while others provide, in addition, an evaluation of the main theses and arguments. Moreover, some synopses are longer than others. If you have to write a synopsis for an assignment, check the assignment instructions for the length requirement and for whether the synopsis should contain an evaluative component.

A useful way to start creating a synopsis is to highlight the main point in each paragraph of the work to be summarized. Review the highlighted points for repetition and redundancy, and disregard repeated or irrelevant points. Try to re-word the remaining highlighted points in your own words, and integrate your sentences into a summary using organizational patterns (see section 9.2).

The following is an example of a purely descriptive synopsis of the practice reading in section 6.3). Compare the synopsis to the reading. As an exercise, try to convert the synopsis to one that contains an evaluative component.

In *The Egg and the Sperm*, Emily Martin is concerned about cultural gender-stereotypes that she claims are used to construct accounts of the way egg and sperm behave. She argues that gender stereotypes in scientific language can harm women, and that one way to help prevent this is to become aware of gender stereotypes in the scientific language of textbooks on human reproduction. She sees herself as contributing to that project in this article.

Martin begins by quoting and discussing examples of the way scientific language in human reproductive biology textbooks tends to describe menstruation in terms that characterize it as dysfunctional, wasteful, and as involving the loss and death of the egg, while sperm production is characterized as powerful and venerable despite the fact that a woman "wastes" only about two-hundred eggs for every baby she produces compared to one-trillion sperm wasted by a man! She goes on to illustrate with examples how scientific language characterizes the egg as a stereotypically helpless, weak female, and the sperm as a stereotypically powerful, active male. To help strengthen her thesis that scientific accounts of human reproduction import cultural gender-stereotypes, she uses examples to illustrate that even when the biological roles of egg and sperm are revised in the light of research showing the egg to be active and the sperm's forward locomotion to be incapable of penetrating the egg, the sperm is still characterized as retaining its active role, only performing it more weakly! The active egg, on the other hand, is described in terms that characterize it as the stereotypical "femme fatale" that 'traps' the sperm. Martin uses the above examples to make a case that scientific accounts of human reproductive biology are not influenced solely by objective facts, but by gender stereotypes in society that express the cultural valuing of males over females.

Martin suggests that the intentionality implicit in cultural stereotypes of gender roles come to be projected on to egg and sperm by scientific language, creating the illusion of the existence of persons at the cellular level: the egg becomes a cellular woman and the sperm becomes a cellular man. She goes on to argue that in consequence, the product of their union, the fertilized egg, is more likely to be regarded as a person also (a cellular baby). Martin suggests that this inaccurate and illusory perspective on the newly fertilized egg could encourage the development of medical technologies to monitor and intervene at earlier and earlier stages of pregnancy on behalf of the baby. Moreover, strong legal rights of the fetus might be legislated, jeopardizing a woman's right to control her own body, including the hard-won right to choose an abortion.

10.3 Writing an Annotated Bibliography

An annotated bibliography is a reference list of sources with notes about the sources after each entry on the list. Notes often take the form of a descriptive synopsis of the source. Depending on the assignment instructions, the synopsis may need to include an evaluative component. A note on the usefulness of the source for one's essay or research project may also be required. Using the source from the previous section, imagine an annotated bibliography assignment that requires a short descriptive synopsis, evaluation, and an indication of the usefulness of the work for one's research. Here is what it could look like:

Martin, E. (1991). The egg and the sperm: How science has constructed a romance based on

stereotypical male-female roles. *Signs: Journal of Women in Culture and Society, 16*(31), 485-

501.

The author presents and discusses numerous examples to support her thesis that scientific accounts of human reproductive biology are constructed from cultural gender-stereotypes. She goes on to argue that the task of exposing the gender-stereotypes in scientific language is necessary in order to disrupt their naturalization, and to prevent them from creating the impression that the newly fertilized egg is a person. This impression, Martin argues, could lead to the loss of women's rights and freedoms in society.

The author's thesis that scientific language constructs accounts of human reproductive biology from cultural gender-stereotypes is well-supported with examples and effective analysis. However, support for the thesis could be strengthened with more empirical evidence, which would help to neutralize the potential criticism that the examples of scientific language used by Martin are anomalous.

This source is useful to my project of criticizing the objectivity of scientific observation because it illustrates how scientific language is biased and subjective. I hope to use some of Martin's examples and analyses to illuminate how scientific protocol sentences might also be laden with cultural stereotypes.

(Annotated bibliographies are useful research tools. When doing research, look for annotated bibliographies on the topics or subject areas you are investigating. You can then sort through large numbers of sources by reading their brief synopses. Many annotated bibliography assignments require you to emulate these professional bibliographies, so check them out.)

10.4 Writing a Case Study

There are many different kinds of case studies, and the term itself means different things to different groups. For the purposes of an academic assignment, a case study often requires students to consider a real life situation for the purpose of attempting to understand it and perhaps, to identify and solve problems, using theories, perspectives or principles studied in the course. In being able to successfully interpret, identify and solve real life situations and their problems from a particular theoretical perspective, cases studies are sometimes regarded as evidence for making generalizations that support the theory used in their analysis.

Organizing Your Case Study Write-Up
Case studies are highly structured documents, often with seven or eight discernible sections. We will briefly review these sections here. The structure of case studies can vary depending on their function, so carefully check the assignment instructions.

1. Synopsis
In this section you should provide a synopsis (see section 10.2) of your case study. Briefly state the purpose of the case study and the theoretical approach or principles used to assess the case. Then provide a concise but brief descriptive overview of the situation that your study focuses on. Close by giving a general statement of the main problems of the case and your proposed solution, without going into specific details.

2. Findings
In this section you should present and discuss the main problems or issues of the case. The problems or issues should be interpreted and evaluated using theories, perspectives or principles studied in the course. The assignment instructions will probably indicate which course concepts you must use, so check the instructions carefully.

3. Discussion
Begin this section by summarizing the main problems identified in the previous section. Then provide a brief, descriptive outline of a range of approaches to solve or help solve the problems.

4. Conclusion
In this section, provide a very brief summary of the main points from the previous two sections.

5. Recommendations and implementation
Choose a solution from the range of approaches presented in the discussion section. Argue for your choice by explaining *how* it will help solve the problems and *why* it is the best solution, with reference to theories, perspectives, or principles from the course. Close by explaining how your proposed solution should be implemented in practical terms.

6. References
Provide a reference list for sources used in your case study.

7. Appendices
Attach any empirical data used but deemed too much to present in the body of the write-up.

10.5 Writing a Scientific Report

The purpose of a scientific report is threefold: 1) to communicate the implications of one's research results, 2) to show how the findings contribute to existing research, and 3) to provide details on how to repeat the experiment at the heart of the research.

Organizing your Scientific Report

A scientific report is a highly structured document with six sections. They are briefly reviewed below. Make sure that the title of your report describes its purpose in detailed, specific terms.

1. Abstract
Provide a concise overview of the entire report. Briefly summarize the hypothesis or question to be tested or answered, the experiment used to test or answer it, the method used to implement the experiment, the results and their implications. (See section 10.1 on writing an abstract.) This section must begin on a new page.

2. Introduction
Contextualize your hypothesis or question by discussing the topic area within which it falls. Explain what has already been established and what still requires investigation. Use this discussion to justify your choice of research hypothesis or question. Present your hypothesis or question clearly and concisely. Be sure to provide in-text citations for all sources referred to. This section must begin on a new page.

3. Methods and Materials
This is perhaps the most important section of the report. Without the possibility of repeating the experiment exactly to verify the results, the other sections of the report are scientifically worthless. Begin by describing the experiment design (i.e. the experiment used to test the hypothesis or answer the question). Then describe in detail the apparatus, method of data collection, and any controls used to implement the experiment design. Describe in the past tense what you did as you conducted the experiment. This section must be detailed enough to enable others to repeat your experiment exactly.

4. Results
In this section, present a clear and concise summary of the results. Summary graphs and tables are acceptable as long as they are actually referred to in the text of this section. Do not evaluate the results. The results section is meant to be purely descriptive.

5. Discussion
Evaluate the results. Explain their implications for your original hypothesis or question. Discuss how your results relate to the work of others in the field. Speculate on the wider significance of your results for the subject area within which your hypothesis or question falls.

6. References
Provide a reference list of all sources used in the report. This section must begin on a new page.

11. APA CITATIONS AND REFERENCES

APA stands for the American Psychological Association, which has developed a writing style for those writing in the discipline of psychology. But the style is also used by people writing in some other areas, including nursing and kinesiology. APA style has been developed primarily for those preparing written work for publication. But as an undergraduate student of kinesiology, your written assignments, although not intended for publication, will be required to include at least some features of APA style. The most common features you will be expected to use in your written work are APA citations and references.

11.1 In-Text Citations

APA requires parenthetical in-text citations. This means that when you directly quote or loosely paraphrase information from a source (no close paraphrase allowed), you must provide bracketed details about the source within the sentences of your paper where you actually quote or loosely paraphrase.

For Direct Quotations

When you give a direct quote you must use the author's exact wording, punctuation and sentence structures. For any quote you include on your paper, you must provide the author's last name, the source's date of publication, and the page number/s of the quoted information, all within brackets, like this:

(Ashby, 2007, p. 79)   (Ashby, 2007, pp. 79-80)

Notice that if the quoted information extends beyond one page in the source, you need to use "pp." rather than just "p." before giving the page numbers.

If the quote is 39 words or fewer, you must place the quoted material within double quotation-marks, and place the citation after the closing quotation-marks but before any end of sentence punctuation. Here are some examples of different ways on which one can give a citation for a quote of 39 words or fewer:

Smith points out that "nurses should never chide terminally ill lung-cancer patients for having been smokers" (2006, p. 50).

It is pointed out that "nurses should never chide terminally ill lung-cancer patients for having been smokers" (Smith, 2006, p. 50).

Smith's point that "nurses should never chide terminally ill lung-cancer patients for having been smokers" (2006, p. 50) addresses the importance of compassionate caring.

Notice that in the first example above, only the date of publication and the page number are given within the brackets, because the author's name has already been given in the signal phrase: "Smith points out" at the beginning of the sentence. In the second example, the author's name is not given in the sentence, so it has to be cited along with the other required details inside the brackets. Note that the period comes after the citation, not before it. The third example illustrates how you can quote and cite in the middle of a sentence.

If the quote is 40 words or more long, it must be presented without quotation-marks as a block of double-spaced sentences, starting on a new line and indented five spaces from the left margin. A citation at the end of a block quote must be placed after any end punctuation. Here are two examples:

> The mind-body problem arises, in part, because it is hard to see how objective, third-person neurophysiological descriptions are conceptually connected to first-person descriptions of conscious thoughts, feelings, and sensations. This rift is a legacy of dualism, the view that a person consists of two different things: a mind, and a body. Anatomy, physiology, and kinesiology have inherited one part of this peculiar dichotomy (the physical) while rejecting the other as not a proper subject for scientific study. (Ashby, 2000, pp.70-71)

Ashby (2000) argues that:

> The mind-body problem arises, in part, because it is hard to see how objective, third-person neurophysiological descriptions are conceptually connected to first-person descriptions of conscious thoughts, feelings, and sensations. This rift is a legacy of dualism, the view that a person consists of two different things: a mind, and a body. Anatomy, physiology, and kinesiology have inherited one part of this peculiar dichotomy (the physical) while rejecting the other as not a proper subject for scientific study. (pp.70-71)

Notice how the citation details are broken up in the second example, the author's name and date of publication appearing within your own sentence at the beginning, and the page numbers appearing after the punctuation at the end of the quote.

For Loose Paraphrase

When you loosely paraphrase, you express an author's idea using your own words and sentence structures. Close paraphrase (using most but not all of the author's own words and sentence structures) is not allowed and is considered a form of cheating (plagiarism). Unlike a quote, a loose paraphrase must not be enclosed by quotation-marks. However, you must provide a citation for a loose paraphrase of whatever length, in the same way as citations are provided for quotations of 39 words or less. The only difference is that with loose paraphrase, you are not compelled to cite page numbers. But if the idea you paraphrase is specific and located on a particular page of the source rather than being a general idea emerging from an entire chapter or whole book, for example, you are encouraged to cite the page number. Here are three examples of loose paraphrase:

Berguer suggests that the almost fanatical obsession with obesity should put us on our guard against body-fascism (2004).

The central idea is that cultural meanings attaching to different body shapes are very unstable and changeable (Wong, 1997).

The argument put forward by Gibolet (2001) shows how to develop a middle way on this issue.

The argument put forward by Gibolet (2001) shows how to develop a middle way on this issue. Gibolet goes on to point out that the dilemma is only apparent, and that this is obvious once we realize that we have been working from erroneous premises.

In the first example above, the brackets contain only the date of publication because the author's name has been given as part of the sentence. In the second example, the author's name is cited inside the brackets along with the date, as the name is not given in the sentence. The third example illustrates that you can cite in the middle of a sentence. Note that citations for loose paraphrase always come before end of sentence punctuation. Also notice that in the fourth example, the date of publication has not been cited in the second sentence. Within the same paragraph, you only need to cite the publication date after the first time if not doing so would create ambiguity about which source the loosely paraphrased information comes from. But with each new paragraph, the date of publication would need to be cited again the first time, even though you had cited it in other paragraphs.

For Sources with Multiple Authors

So far, the examples we have looked at have all been citations of single-authored sources. But what if the source you are loosely paraphrasing or quoting from has been written by more than one author? APA has laid down the following rules.

Sources with Two Authors

If a source you are quoting or loosely paraphrasing has two authors, you must always cite both authors' last names, joined together with the ampersand symbol, like this:

(Smith & Qureshi, 2000, p. 17)   (Petrie & Jones 1997, pp. 3-4)   (Obedu & Tsang, 1995)

Here are two examples:

It is argued that "processes of empowerment through gender-construction are background conditions of social and political systems" (Wressel & Thwaites, 1994, p. 41).

Wressel and Thwaites argue that "processes of empowerment through gender-construction are background conditions of social and political systems" (1994, p. 41).

Notice that in the second example, where the authors' names are given as part of the sentence, the names are not joined by the ampersand symbol but by the word "and." This is because the ampersand symbol is not a proper grammatical part of a sentence and must only be used within the citation brackets.

Sources with Three to Five Authors

If a source you are loosely paraphrasing or quoting has three, four, or five authors, you must cite all their last names the first time. On subsequent occasions *throughout your paper*, cite the last name of the first author, followed by the phrase 'et al.' This phrase must not be in italics and there must be a period after the 'al'. Here is how all this works:

First Time	Subsequent Occasions
(Yolens, Aris & Logie, 2002, p. 61)	(Yolens et al., 2002, p.61)
(Maillardet, Poy, Gill & Beeg, 2003, pp. 40-41)	(Maillardet et al. 2003, pp.40-41)
(Singh, Micot, Koch, Recouz & Rau, 2004)	(Singh et al., 2004)

Notice that in the 'first time' column, in each case the last two names are joined by the ampersand and former names by commas. Also note that on subsequent occasions *within the same paragraph*, the citations in the 'subsequent occasions' column might be even shorter because you would be permitted to omit the date of publication if this caused no ambiguity about which source the paraphrase or quote was from.

Sources with Six or more Authors

If a source you are loosely paraphrasing or quoting has six or more authors, only cite the last name of the first author, followed by the phrase: 'et al.' However, unlike citations for sources with three to five authors, you must always include the date of publication in subsequent citations of a source with six or more authors. Here is how this works for a source authored by Viollier, Bidlake, Hisette, Pohl, Foix, and Duhamel (six authors):

First Time	All Subsequent Occasions
(Viollier et al., 1990)	(Viollier et al., 1990)

Note that for citations of a source with six or more authors, citations on subsequent occasions are the same as the first time the source is cited.

For Work Cited within Secondary Sources

If source A quotes or paraphrases information from another source B, and you paraphrase A's quote or paraphrase of B, you need to indicate that you are essentially paraphrasing information from source B. You do this by using the phrase "as cited in." For your paraphrase of B's information, you would cite like this: (B, as cited in A). Let's take an example. Suppose you use a source authored by Suzanne Lazenby and published in 1990. Lazenby quotes information from a source published in 1988 and authored by Nelson Grahn. You have selected the Lazenby source for your paper and you paraphrase Lazenby's quote from Grahn. Here's how you would cite:

(Grahn, as cited in Lazenby, 1990)

You would list the Lazenby source but not the Grahn source on your references page. If you wished to use the same direct quotation from Grahn as the one used by Lazenby, it would be proper research procedure to locate the source by Grahn, read the quote in context and cite Grahn as a separate source, independently of Lazenby. In this case, you would need to list Grahn as a source on your references page.

For Multiple Sources

Suppose you paraphrased a point made in more than one source. You would then need to cite multiple sources for this paraphrase. How would you do it? You could either give the authors' names as part of the grammatical sentence, joined by conjunction words such as "and," or you could give multiple citations, separated by semicolons, within the brackets. These alternatives are illustrated by the following two examples:

Hoffstead (1990), Broglie et al. (1994), and Patel and Mudge (1996) point out that more funding for physical education will mean greater savings to the healthcare system.

More money needs to be put into physical education as this will mean greater savings to the healthcare system (Hoffstead, 1990; Broglie et al., 1994; Patel & Mudge, 1996).

For Personal Lecture Notes

Personal lecture notes are notes that you have made for yourself rather than copied from overheads or PowerPoint slides. You would cite personal lecture notes by giving the name of the lecturer, the course code, and the date of the lecture. Here are two examples:

Professor William Peplow, in a KINE1000 lecture on November 15, 2004 argued that the upper limits of optimum athletic performance have already been reached, and that further advancement depends on sports medicine.

Professor William Peplow argued that the upper limits of optimum athletic performance have already been reached, and that further advancement depends on sports medicine (KINE1000 lecture, November 15, 2004).

Note that you should not list your personal lecture notes as sources on your references page, as they are regarded as sources of non-recoverable information.

For Lecture Notes Posted on WebCT

Lecture notes that are printed and distributed or posted on Webct or on another form of course homepage are regarded as a published source and must be treated as such when you quote or paraphrase them. You would cite them in the same way as you would cite any published, authored source – with the author's last name and the date of publication. For example:

(Peplow, 2004)

If you quote from posted lecture notes, you may not be able to give a page number if the notes are not paginated. But, try to be as specific as possible by citing the exact date if the notes are dated. If not, cite the title of the notes if there is one. As a source of recoverable information, printed or posted lecture note sources must be listed on your references page.

For Lecture PowerPoint Presentations

PowerPoint lecture slides should be cited in the same way as printed or posted lecture notes (see above). But you should give the slide number instead of a page number if you quote from a PowerPoint Presentation. As a source of recoverable information, if you use details from a lecture PowerPoint presentation in your paper, you must list the presentation on your references page.

For Course Readers

Course readers bring together a collection of authored readings. It is the individual readings in the course reader that you would use for a paper, so it is these that you would cite. Readings in a course reader should be treated as reprinted sources, except for any readings that have been purposely written for the reader. To cite a reading from a course reader, give the last name of the author of the reading, the original date of publication followed by a slash, the date of the course reader, and then page numbers if you are quoting from the reading. Take an example. Suppose you are in a course that uses a course reader which was put together in 2001. You wish to use a reading from the course reader, and the reading is authored by Jacqueline Simons and originally published in 1996. Here are two examples of how you might cite a quote or paraphrase from this reading in the course reader:

(Simons, 1996/2001)   (Simons, 1996/2001, p. 24)

Note that on your references page, you would need to list the course reader along with the original publication details, not just the original publication details, because you have cited the source as a reprint and so need to list the reprint details (i.e. the course reader).

For Course Manuals

Course manuals do not bring collections of authored readings together, but instead, present information, skills, techniques and principles that are important and relevant to the course. Course manuals are often authored by lecturers of the course they are intended for. You are reading a course manual at this moment! You would cite quotations or paraphrases from course manuals in the usual way, giving the last name of the author, the date of publication, and the page number if you have quoted from the manual. Course manuals are sources of recoverable information, so they must be listed on your references page.

For Class Handouts

Class handouts should be cited in the same way as for printed or posted lecture notes. Give the last name of the author (the lecturer), the date, and the page number (if there is one) if you are quoting from the handout. Again, as a source of recoverable information, if you paraphrase or quote from a course handout, you must list the handout on your references page.

For Internet Sources

Internet sources should be cited in the same way as other kinds of sources, using the author-date style, as above. However, Internet sources frequently present problems that force one to deviate from the author-date citation formula. Let's take a look at some of these problems and how they are dealt with.

Sources with No Author

Sometimes, a source does not indicate an author. This is quite a common problem with Internet sources. Be careful, however, to make sure that you have not overlooked the author. The author may not be a human being, but an organization or institute, in which case the source might indicate as author the name of an organization or institute rather than a human name. Check this out before concluding that the source indicates no author. If the source indicates no author, you must cite the first few words of the title in double quotation-marks (if it is the title of an article, or a chapter or section of a larger work such as a book or Web document) or in italics (if it is the title of an entire report, book, periodical, or non-periodical Web document). Here are two examples:

("Role of Carbohydrates," 1990)   (*Athletics and Critical Skills*, 2002, p. 10)

Sources with No Date of Publication

Sometimes, a source does not indicate a date of publication. In such cases, you need to write "n.d." (which means 'no date') where the date would go in the citation. Here's how it looks:

(Chou, n.d.)   (Syed & Kaul, n.d., pp. 67-68)

Sources with No Page Numbers

Internet sources are often not paginated. Yet APA style requires you to give page numbers when you quote, and preferably when you paraphrase too. So what are you to do! If the paragraphs of the source are numbered, simply give the number of the paragraph where the quote comes from. In your citation, either place the paragraph symbol ¶ before the number or write "para." before the number. Be consistent in your choice. If you use the symbol, always use the symbol in such cases. If you use the word, always use the word. This is how it looks:

(Nycod, 1985, ¶ 3)   (Nycod, 1985, para. 3)

Sometimes, a source not only has no page numbers, but the paragraphs are not numbered either. In this case, cite the title of the relevant section-heading and count the number of paragraphs from the beginning of this section to the paragraph your quote comes from. Cite the number of this paragraph. This is how it looks:

(Sowter, 2005, Conclusion section, ¶ 7)   (Sowter, 2005, Conclusion section, para. 7)

Sources with Multiple Problems

It is quite possible that you will come across sources that pose several of the problems that we have just been looking at. For example, a source might indicate neither an author nor a publication date, and have no page numbers or numbered paragraphs. On these occasions, you would use more than one of the above conventions in your citation. The example below illustrates how you would cite a worst case scenario source!

("Tough Love," n.d., Alternative Models section, ¶ 11)

This is an example of a citation for a source with no indicated author, no indicated date of publication, no page numbers, and no numbered paragraphs. It doesn't get much worse than that!

Web site or Document on Web site?

Students sometimes confuse documents on Web sites with the Web sites themselves. APA encourages you to cite and reference documents on Web sites rather than citing entire Web sites. When you cite an entire Web site, you must simply cite the Web site's URL in the text of your paper inside parentheses, but not list the Web site or its URL on your references page. Here is an example:

The Women's Sports Foundation (http://www.womenssportsfoundation.org/cgi-bin/iowa/ index.html) provides useful information, articles, updates and links on all aspects of women's involvement in sports and athletics, and much more.

Notice that the URL has been broken up at a slash. APA stipulates that URLs may be broken up immediately after slashes or immediately before periods. Never add punctuation of your own at the place where you break up a URL.

When you cite a document on a Web site, you must look for an author and date of publication and follow the author-date formula. Do not cite the URL in the text of your paper. On your references page you would list the source, giving the URL that takes the reader to the document, not to the Web site's home page.

11.2 Reference Lists

The purpose of an in-text citation is to provide brief details of a source (author, and date of publication) so as not to distract the reader from what s/he is reading. But an in-text citation is rich enough in information to allow the reader to cross-check it with full details of the source. The full details of sources are listed on a references page at the end of the paper. A references page is not a bibliography. All sources listed on a references page must have been cited in the text of your paper. The APA conventions governing the style of reference lists are there to ensure that the reader can cross-check text citations with entries on the references page quickly and easily, without having to ferret about for the relevant entry. Below, the basic APA format requirements for references lists are set out:

- **Reference list on a separate page.** Your reference list must be on a separate page right after the text of your paper and before any attachments, such as appendices

- **Four basic elements.** Each source entry on your reference list must include the author's name, date of publication, title of source, and publication details

- **Double line-spacing.** Your reference list should be double-spaced throughout. Alternatively, you may double-space between single-spaced source entries

- **Indent subsequent lines of entries.** The first line of a source entry must appear immediately against the left margin. Subsequent lines of a source entry must be indented three spaces

- **Source entries must be arranged alphabetically by authors' last names.** For sources that have no author, use the first letter of the title to guide you as to where the source entry should be placed

- **"References."** Give your references list the title: "References," not "Bibliography."

Tip: Make sure that you do not change the order in which authors' names appear. The order in which names appear *within an entry* on your references page must correspond exactly to the order in which names are cited in text for sources with multiple authors. Otherwise, the reader cannot quickly and easily cross-check in-text citations with entries on the references page! **The information cited in text must correspond exactly with the information that appears immediately against the left margin on your references page.**

In addition to providing conventions for formatting the reference list, APA has developed style formulae for how information for different types of sources must be entered on the reference list. A convention that applies to all types of sources is that the names of authors up to a maximum of six must be listed. For sources with more than six authors, the first six are listed followed by the phrase "et al." Here is what this convention looks like in practice:

Author, I.	One author
Author, I., & Author, I.	Two authors
Author, I., Author, I., & Author, I.	Three authors
Author, I., Author, I., Author, I., & Author, I.	Four authors
Author, I., Author, I., Author, I., Author, I., & Author, I.	Five authors
Author, I., Author, I., Author, I., Author, I., Author, I., Author, I.	Six authors
Author, I., Author, I., Author, I., Author, I., Author, I., Author, I. et al.	Seven + authors

Notice that this is not what you do when you cite a source in text! The phrase "et al." is used after the last name of the first author of a source with six or more authors that is cited in the text. We will now review some of the more common types of sources that you are likely to come across.

1) <u>Books</u>

Authored Books

Chapters of an authored book are all written by the same author or authors. Publication details include the city, and abbreviated state or province, and the country if the publisher is located outside of the United States. However, if the publisher is located in the following places, it is not necessary to include abbreviated state, province or country: London, New York, Jerusalem, Stockholm, Chicago, Moscow, Philadelphia, Rome, Tokyo, Los Angeles, Vienna, Baltimore, Milan, San Francisco, and Boston. The words "Co.," "Inc.," and "Company" are omitted from the names of publishing houses. Note that book titles are in italics, with lower case except for the first letter of the first word of the title, and the first letter of title words that are proper adjectives or proper pronouns. If the edition of the book is indicated, give this information in brackets after the book title and before the period.

Formula:

Author, I. (Date of publication). *Title of book*. City, State: Publisher.

Example:

Coakley, J., & Donnelly, P. (2004). *Sports in society: Issues and controversies* (1st Canadian Ed.). Toronto: ON. McGraw-Hill Ryerson.

Chapter from an Authored Book

If only a particular chapter of an authored book is relevant to your paper, you should give a reference list entry for the chapter, not for the whole book. Give the page numbers for the chapter in brackets after the title and before the period.

Formula:

Author, I. (Date of publication). Title of chapter. In *Title of book* (pp. page numbers for the chapter). City, State: Publisher.

Example:

Coakley, J., & Donnelly, P. (2004). Using social theories: What can they tell us about sports and society? In *Sports in society: Issues and controversies* (1st Canadian ed., pp.30-53). Toronto, ON: McGraw-Hill Ryerson.

Edited Books

Authored books can have editors, who write comments in an introduction about the book and the author, and who might have had a hand in preparing the manuscript for republication. But there is another group of books, the chapters of which are written by different authors. This is what we mean be edited books here – the editors have often been involved in selecting different authors' work for the chapters of the book. Editors of such books must be indicated as such, giving (Ed.) or (Eds.) in brackets after their names.

Formula:

Editor, I. (Ed.). (Date of publication). *Title of book*. City, State: Publisher.

Example:

Curtis, J., & Russell, S. (Eds.). (1997). *Physical activity in human experience: Interdisciplinary perspectives*. Champaign, IL: Human Kinetics.

Chapter from an Edited Book

If only a particular chapter of an edited book is relevant to your paper, you should give a reference list entry for the chapter, not for the whole book. Notice that you must give the editors names after "In" and before the title of the book as a whole. Notice also that their initials come before their last names. Give the page numbers for the chapter in brackets after the title and before the period. This is certainly one of the more tricky reference list formulae to follow and get right, so pay close attention!

Formula:

Author, I. (Date of publication). Title of chapter. In I. Editor (Ed.), *Title of book* (pp. page numbers for chapter). City, State: Publisher.

Example:

Gruneau, R. (1997). The politics and ideology of active living in historical perspective. In J. Curtis & S. Russell (Eds.), *Physical activity in human experience: Interdisciplinary perspectives* (pp. 191-228). Champaign, IL: Human Kinetics.

2) Journal Articles

Reference list entries for journal articles must give the title of the article, the title of the journal the article is from, the volume and issue numbers, and the pages. Notice that the words "volume," "vol.," and "issue" are not used – simply give the volume number, and then the issue number next to it inside brackets. The journal title, unlike the article title, is in upper case. The italics include the volume number. Notice also that for journal articles, you do not write "pp." before giving the article page numbers.

Formula:

Author, I. (Date of publication). Title of article. *Journal Title, volume number*(issue number),

page numbers for article.

Example:

Hastings, D., Kurth, S., & Schloder, M. (1996). Work routines in the serious careers of Canadian

and U.S. masters swimmers. *Avante, 2*(3), 73-92.

3) Book Reviews

Book reviews can be useful sources of critical review. But be aware that a book review, at the back of an academic journal, is not an academic journal article!

Formula:

Author, I. (Date of publication). Title of review. [Review of the book *Title of book*]. *Title of*

Source of Book Review, volume number, page.

Example:

Markula, P. (1999). [Review of the book *Strong women, deep closets*]. *Sociology of Sport*

Journal, 16(2), 174-176.

Notice in the above example that no title of the review has been given. This is because there isn't one in this particular case.

4) Daily Newspaper Articles

Unless the assignment instructions state otherwise, you don't want to base your research essay on too many of these! But one or two topical articles may do no harm. Remember to give the full date of publication.

Formula:

Author, I. (Date of publication). Article title. *Newspaper Title*, p. page number for the article.

Examples:

Ogilvie, M. (2006, November 22). Unequal city. *Toronto Star*, pp. B1, B4.

Residents' health tied to neighbourhoods. (2006, November 22). *Toronto Star*, pp. A1, A22.

Notice that the article in the second example has no author. This is why the article title is given first, followed by the date. Notice also that in both examples, the article is spread over discontinuous pages.

5) Magazine Articles

Again, you don't want to base your research essay on too many of these. But there is no reason why you should not use one or two, unless the assignment instructions state otherwise! Be careful to give the year and month for monthly magazines, and the year, month and day for weekly magazines. Notice that unlike newspaper articles, you do not write "p." or "pp." before giving the article page numbers.

Formula:

Author, I. (Date of publication). Title of article. *Title of Magazine, volume*, page numbers.

Examples:

Andersen, J. L., Schjerling, P., & Saltin, B. (2000, September). Muscles, genes and athletic performance. *Scientific American, 283*, 48-55.

The politics of genes: America's next ethical war. (2001, April 14). *The Economist, 359*, 21-24.

Notice that the article in the second example has no author. This is why the article title is given first, followed by the date.

INTERNET SOURCES

The APA conventions of style above apply also to Internet sources. But you need to provide additional information, because Internet sources are located on Web sites, and Web sites are frequently updated and sometimes deleted! Hence, in addition to information as above, you also usually need to give the date of retrieval for when you actually accessed the source, as well as the URL address that will take the reader to the source. You must also give a Web site descriptor before giving the URL if the source comes from a large and complex Web site, such as a government agency or university Web site. We will now review some of the more common Internet sources that you are likely to come across.

6) Internet-Only Newsletters

These are often disseminated by email. They consist of news with updates, developments, and sometimes articles within the specialist fields that they serve. Give the full date of publication, as indicated in the newsletter. Notice that there is no end punctuation after a URL address. APA states that a URL may be broken up after a slash or before a period.

Formula:

Author, I. (Date of publication). Article title. *Newsletter title, volume(issue).* Retrieved Date of

retrieval, from URL address

Example:

Holt, J. (2004, November 10). The rise of pairing-based cryptography and identity-based

encryption. *Cipher, E63.* Retrieved June 5, 2007, from http://www.ieee-security.org/

Cipher/Newsbriefs/2004/111804.ListWatch.html#ecc

7) Articles from Internet-Only Journals

There has been a rise in the number of e-journals in recent years. E-journals sometimes do not use page numbers for their articles and instead, simply number the articles. On your reference list, you must use the system that the e-journal employs, whether it is page numbering or something else. Check the peer-review processes of e-journals.

Formula:

Author, I. (Date of publication). Article title. *Journal Title, volume,* page numbers or other

article designator. Retrieved Date of retrieval, from URL address

Example:

Zeitz, K., Kadow-Griffin, C., & Zeitz, C. (2005). Injury occurrences at a mass gathering event. *Journal of Emergency Primary Health Care, 3,* Article 990098. Retrieved May 30, 2007, from http://www.jephc.com/full_article.cfm?content_id=207

8) Articles from Journals that Publish in Print Form and on the Internet

Many journals publish their issues in print form and electronically on the Internet. If you access an article from such a journal via the Internet, you should let the reader know with the phrase: "Electronic version." Give the date of retrieval and the URL address. Technically, you do not have to indicate that it is an electronic version, give a date of retrieval or a URL address if you have no reason to think that the electronic version uses a different format or page numbering to the print version. However, if you are not sure, it may be wise to err on the side of caution. The following formula and example reflect this choice.

Formula:

Author, I. (Date). Article title [Electronic version.]. *Journal Title, volume*, pages. Retrieved Date of retrieval, from URL address

Example:

Vonderhaar, J. M., & Campbell, B. M. (2005). A model for delivering exercise interventions to address overweight and obesity in adults: Recommendations from the American Kinesiotherapy Association [Electronic version.]. *Clinical Kinesiology, 59,* 39-42. Retrieved May 20, 2007, from http://www.clinicalkinesiology.org/

9) Report on a Government Department Web site

For government reports, the author is often the name of a commission or advisory body.

Formula:

Author, I. (Date of publication). *Title of report.* Retrieved Date of retrieval, from Web site descriptor: URL address

Example:

Commission on the Future of Health Care in Canada. (2002, November 28). *Final report: Building on values: The future of health care in Canada.* Retrieved June 6, 2007, from Health Canada Web site: http://www.hc-sc.gc.ca/English/pdf/romanow/pdfs/HCC_Final_Report.pdf

Notice that this example uses the Web site descriptor: "Health Canada Web site:" before giving the URL address.

10) Document available on a Government Agency Web site

This is similar to the above. The agency itself is usually the official author of the document, unless indicated otherwise. Notice the Web site descriptor in the example.

Formula:

Author, I. (Date of publication). *Title of document.* Retrieved Date of Retrieval, from Web site descriptor: URL address

Example:

Public Health Agency of Canada. (2005, April 14). *Population health approach.* Retrieved June 7, 2007, from Public Health Agency of Canada Web site: http://www.phac-aspc.gc.ca/ph-sp/phdd/index.html

11) Course Reader

As was mentioned earlier in section 6.1, a course reader consists of the lecturer's selection of different authored sources. You should not give the course reader in its entirety as an entry on your reference list. Instead, you should list the relevant source from within it. The source will need to be treated as if it is reprinted in an edited book, with the lecturer as the editor.

Formula:

Author, I. (Date of recent re-publication). Title of source. In I. Editor (Ed.), *Title of course reader* (pp. page numbers for source). City, Province: Name of University, Department of School. (Reprinted from *Title of Original Source* of the re-printed source, volume(issue), original pages, original date of publication)

Example:

Martin, E. (2006). The egg and the sperm: How science has constructed a romance based on stereotypical male-female roles. In N. Ashby (Ed.), *KINE1000 Sociocultural perspectives in kinesiology course reader* (pp. 41-52). Toronto, ON: York University, School of Kinesiology. (Reprinted from *Signs: Journal of Women in Culture and Society, 16*(31), 485-501, 1991)

12) Course Manual

A course manual contains information, skills, and techniques authored by the lecturer and relevant to the course. Hence, course manuals should be entered on your reference list in the same way as an authored book, with the lecturer as the author and the university as the publisher, unless otherwise indicated.

Formula:

Author, I. (Date of publication). *Title of manual*. City, Province: Publisher.

Example:

Ashby, N. (2006). *KINE1000 Sociocultural perspectives in kinesiology critical and writing skills manual*. Toronto, ON: York University.

13) Lecture PowerPoint Presentation

PowerPoint presentations are often used by lecturers as their lecture notes. If not posted on Webct they can be entered on your reference list using the unpublished paper presented at a meeting formula.

Formula:

Author, I. (Date of presentation). *Title of PowerPoint presentation*. PowerPoint presentation at Give details of where the presentation took place, City, Province.

Example:

Ashby, N. (2006, Monday 18). *Critical skills for active reading*. PowerPoint presentation at a KINE1000 lecture at York University, Toronto, ON.

If the lecture PowerPoint presentation is posted on Webct, the 'report from a university' is probably the better choice of formula with which to list it.

Formula:

Author, I. (Date of publication). *Title of PowerPoint presentation* (PowerPoint Presentation).

City, Province: Name of University, Department.

Example:

Humaña, H. (2006). *Overview of mid-term exam* (PowerPoint Presentation). Toronto, ON: York

University, School of Kinesiology, KINE1000. Retrieved December 4, 2006, from

http://webct.yorku.ca/SCRIPT/2006_AS_KINE_Y_1000_6_A_EN_A_LECT_01/scripts/

serve_home

When the PowerPoint presentation has been posted on Webct, you need to give a date of retrieval and an address, as above.

11.3 Resources

In this short manual, it has not been possible to address many of the nuances and complexities of APA in-text citations and reference lists. You are encouraged to go on to explore these for yourself. Here are some resources that you may find useful.

- The most important resource to become acquainted with is the American Psychological Association's own published style guide, the *Publication Manual of the American Psychological Association*. You will need the latest edition, which at this time is the fifth edition, published in 2001. University libraries will have it. Call number: BF 76.7 P83

- Check out APA Online at http://apastyle.apa.org/ Here you can access information, tips, and look at responses to frequently asked questions about APA style

- Access and use Purdue University's Online Writing Lab (OWL), which provides information on APA formatting, in-text citations, and reference lists, as well as an APA overview and workshop. Access OWL at http://owl.english.purdue.edu/owl/

- Check out the Citation Machine at http://citationmachine.net/ However, the software has some flaws. It is suggested that you use the Citation Machine simply as an educational tool to explore APA style and not as a tool to create in-text citations and reference lists that you would actually include in your papers.

12. CLASS PRESENTATION SKILLS

The art of giving a talk at university has a long history, dating back to universities of the Middle Ages, when students were required to give a successful, learned and persuasive speech, called a *Disputation*, in front of peers and public in order to receive a degree. There have always been vocal presentation assignments at university, but they seem to be becoming even more popular as a way of combating cheating and plagiarism – it is hard to give a convincing vocal presentation, including a questions and answers session, if you don't really understand the topic you are talking about! Here, we will review the basic skills involved in giving a vocal presentation, with some focus on the use of PowerPoint, which has become the most popular support aid.

12.1 Researching your Topic and Writing the Script

The nature of your research depends largely on the assignment instructions. It may be that you are required to research a topic in depth from a list or of your own choosing, and then give a twenty minutes vocal presentation on the topic. If so, use the research skills set out in section 7 of this manual. Alternatively, you may be required to select a reading from your course reader and give a ten minutes presentation on it. In this case, for your research use the active reading skills set out in section 6 of this manual.

Once you have completed your research and evaluated your source or sources, it is time to draft the script of your presentation using the writing process skills discussed in section 8 of this manual. Begin by creating a draft outline of your presentation in mind map or point form, and amend it as necessary as you proceed through the draft writing stages – the outline will prove useful later on. The structure of your presentation will be similar to that of an essay, with 1) an opening introduction, in which you indicate the topic, give a brief overview of how the presentation will unfold, and state your view, 2) a body, in which you discuss and evaluate main ideas, facts, evidence, arguments, or issues, and 3) a conclusion, in which you summarize key points, repeat your thesis, and bring the talk to a close. Unlike an essay, however, where you are writing primarily for a captive audience (your professor or TA who must read and grade it), with a presentation script you are writing for an audience that can tune out if it is unclear or totally boring. Here are some tips as you write the first draft of your presentation script:

> - **Capture the audience's attention in the opening remarks of your introduction.** A common piece of advice to essay writers is to capture the attention of the reader at the start, with a witty or appropriate aphorism, anecdote, or surprising or shocking fact related to the essay topic. The same advice applies to the opening introduction of a presentation. First impressions are important, and the audience will be more reluctant to tune out if you capture their attention right at the start. So incorporate something interesting, surprising or shocking into the opening sentences of your presentation script

> - **Be clear and concise in your introduction.** After you have captured your audience's attention, you want to keep it. State clearly and simply what your presentation is about, why you think it is important (your thesis), and very briefly indicate how your presentation will unfold. Mention main points. Do not ramble

- **Don't allow details to cloud main points in the body.** In the body of your script, you want to introduce main points, and discuss and evaluate them. But it is important that details of the discussion and evaluation do not go on and on without returning to the main point. Listeners can only keep a little in their minds at a time. If you go into a lengthy discussion of details, after a short while your audience will forget which main point the discussion is connected to! Many will tune out. So, as you discuss and evaluate, arrange it so that each strand of discussion often returns to the main point to sum up what you have shown so far. Keep the connection between main points and details clear

- **Employ appropriate organizational patterns in the body of your script.** You need to sequence the main points and details in such a way that they follow each other in the clearest way. If the audience feels lost or has no sense of the direction of your talk, they will tune out. Use organizational patterns (see section 9.2) to arrange your points so that the transition from one point to the next is as clear and logical as possible. Use indicator words and phrases, such as "we now turn to..." to signal transitions and maintain a sense of the direction of your presentation for the audience

- **Write pauses into the body.** Remember that you are writing the script of a vocal presentation, not an essay. Anticipate passages in your discussion and evaluation that the audience may find heavy-going even though the passages are clear. Use brackets and write "Pause" after challenging passages of your script, like this: [PAUSE]

- **Summarize key points in your conclusion.** Do not summarize all of your main points. The audience will be tired from mental effort toward the end of your presentation. Repeat only those points that are absolutely key to supporting your overall message or thesis. Keep the conclusion short. Do not ramble

- **Capture the audience's attention in the closing remarks of your conclusion.** Final impressions are as important as first impressions. You want your audience to remember your presentation as a worthwhile experience, so end with something interesting. If you started the presentation with a witty or appropriate aphorism, it would be nice to end with one. Alternatively, underscore your overall message or thesis by connecting it to a surprising, important, or serious actual or possible situation or development in society. Don't go off at a tangent and begin a new talk – you are only mentioning it *very briefly* as a way of bringing your presentation to a close in a memorable way.

Once you have written the first draft, put it aside for a day or two. When you come back to it, you need to experience yourself giving the presentation and hearing your script in order to be able to make appropriate second draft changes. Record and time yourself reading the script out loud. Make changes to any passages that sound too heavy, convoluted, or repetitious. If your reading finishes way too soon or far too late relative to the time allowed for the presentation, listen to the recording to check that you are not speaking too fast or too slowly. Then make cuts or additions to the script if necessary.

12.2 Preparing to Give your Presentation

When the second draft of your script is complete, it is time to practice giving the presentation. To begin with, record yourself as you read out your script to willing family members and friends. Don't forget to pause for about ten seconds at places in the script where you have indicated a pause. Take note of your practice audience's observations and criticisms and make appropriate changes to the content of the script or to your delivery of it. Listen to the recordings and pay attention to the way you sound. Are you reading too quickly or too slowly? Are you hesitating between points too much? Does your voice sound monotone? Are you speaking so quietly that it is hard to hear what you're saying? Do you sound confident, uncertain, or anxious? Try again, correcting these flaws and varying the intonation of your voice as you introduce new ideas or make important inferences. A monotone voice would make your audience tune out. You need to develop a 'sing-song' up and down tone as you deliver your presentation – but obviously, don't exaggerate a varying intonation to the point where it sounds contrived.

As you practice, you should also pay attention to what you are doing with your body. Are you just standing on the spot? Are your arms by your sides? Is your head up or bowed? Body-language is important when you are on display to an audience. As you give your presentation, practice pacing very slowly from the middle of your space to the left or right, and then back again. Giving the audience a moving target keeps them interested. Use your arms to help your intonation underscore important points. For example, raise your left or right arm to eye level, put thumb and first two forefingers together, and prick the air in front of you as you make an important observation. If you want to express the magnitude of an issue or idea as you introduce it vocally, stretch your arms out wide in front of you with your palms facing inwards, as if you are holding a large package. Another way of emphasizing an important point is to step slightly closer to the audience as you make the point. If you can, have someone use a camcorder as you practice. View the results and make changes until your body-language works like your voice to animate your presentation without looking contrived.

As you practice, you will start to memorize your script and begin to be able to give your presentation without it. Try using the point form or mind map script outline that you created at the draft stage. With enough practice, you will find that you are able to improvise your vocal presentation around the script outline, with only the occasional glance at the outline. Practice giving your improvised presentation while maintaining eye contact with your audience and without often looking at the script outline. This is important, because you will need to keep eye contact with your audience when you give the presentation for real. Summing up:

> - **Record yourself reading the script out loud to willing family members and friends**
> - **Take note of your practice audience's observations and criticisms**
> - **Pay attention to the way you sound**
> - **Pay attention to the way you use your body**
> - **Memorize your script**
> - **Practice giving your presentation with the script outline alone**
> - **Practice improvising your presentation with just an occasional glance at the script outline**

12.3 On the Day of your Presentation

You have practiced your presentation and style of delivery, you are all ready to go, and the big day has arrived! The chief feeling running through your body is likely anxiety and butterflies in the tummy. This is perfectly normal and not something to become further worried about. Just accept that standing in front of a group of people and giving a talk is one of those activities that naturally create some nervousness even in the most seasoned public speakers. Moreover, remind yourself that you have, by and large, a friendly audience. Course instructors and TAs want you to succeed, and they will ensure that you do not have a bad experience. Your fellow students will likely have to do the same assignment, with you in the audience. They, too, will greet your presentation with enthusiasm because they will hope for the same support from you when their time comes. Hence, accept your anxiety but at the same time, recognize that fear of your audience is groundless. Turn your nervousness into something positive – focus it on your desire to give a great performance, just as you have done many times during practice. Let your anxiety help you forget everything except giving a perfect presentation – not because something terrible would happen if you did not, but because you deserve to be recognized for all the hard work you have put into it.

Bring your script outline with you, whether in point form or as a mind map. Make sure that it is large enough to be seen easily if you get lost in your talk and need to refer to it. Also bring your full script with you. You will not use it, but the knowledge that it is with you can be a huge stress reliever – it helps to neutralize the fantasy worst-case scenario of being at a loss for words in front of a huge audience – you'd have the script to fall back on. Bring a watch or small pocket clock with you – something with an easy to see display of the time. This will help you keep track of time as you give the presentation. Finally, bring some bottled water in case you get a dry throat.

Arrive fifteen or twenty minutes early, if possible. The venue is likely a classroom or lecture hall. Greet students as they arrive. Chatting and joking with them in a friendly way about how you feel before your presentation starts is a real anxiety reducer. If it is a large venue, it is probably wired for sound. If s/he has not already done so, ask the instructor or TA to set up the audio equipment for you and then check it out. Make sure that the volume is sufficiently turned up so that people at the back can hear, and so that you do not have to shout or talk so loudly that it is not possible to vary the intonation of your voice. Place your script outline in a suitable location at the front where you will be giving the presentation, so that you can easily glance at it if necessary. Place your watch or clock on a convenient table or on the podium where you can glance at it surreptitiously. Don't wear the watch and keep looking at it - that would come across as rude.
Notice how much space you have and think about how you will move around in this space. If it is a classroom or lecture hall used by the course, you will probably already have some familiarity with the layout.

When it is time for your presentation to begin, introduce yourself and say that you are pleased to have the opportunity to talk about the topic. If you are entirely new to giving a presentation, and there is supposed to be a questions and answers session, state that members of the audience will be given a chance to ask questions at the end. It is probably not a good idea to invite or take questions after each main section of your presentation, as it may be difficult to manage the discussion and you might lose control of the presentation. Far better to have the questions and answers session at the end until you become a presentations pro. Once you have introduced yourself and indicated when audience participation will be allowed, go into your presentation. Remember the points about voice delivery and body-language. Maintain eye contact with the audience, vary your intonation, and use body-language to get your points across. Don't forget those pauses! Secretly glance at your watch or clock from time to time, to ensure that you are roughly where you need to be in your talk. Speed your delivery up or slow it down just a little, if necessary. If you receive any unplanned questions from members of the audience, thank then for their interesting questions and say that you will be happy to address them at the end.

During the questions and answers session, if there is one, you will likely receive a rag-bag of questions, some excellent, and some revealing misunderstandings. Repeat questions that you receive so that everyone can hear them and to confirm that you have understood them correctly. Receive all questions with courtesy. If you don't know the answer to a question, say so, and thank the member of the audience for an interesting point that you will reflect on. Do not waffle in response to a question you don't know the answer to – this would make it seem as if you do not know what you are talking about! If someone asks an unclear question, ask for it to be rephrased, or rephrase it yourself out loud and seek acknowledgement from the questioner that you have understood the question. When you receive a question that reveals misunderstanding on the part of the questioner, try to convert the question into one that would be a great question to ask, and then say that if the questioner meant this question, you would answer in such and such a way.

Summing up:

- **Focus anxiety on the task of giving a great presentation, not on fear of the audience**
- **Bring your script outline to glance at if you forget your direction in the talk**
- **Bring the full script as a psychological aid**
- **Bring a watch or clock to keep track of time**
- **Bring bottled water in case of a dry throat**
- **Arrive early to check the venue, the sound system, and to chat and joke with members of the audience before the presentation begins**
- **At the beginning of the presentation, state that you will be happy to receive questions at the end**
- **Remember to use vocal intonation and body-language**
- **Check the time occasionally, and speed up or slow down your delivery a little if you get behind or ahead of schedule**
- **Receive all questions courteously**
- **If you don't know the answer to a question, say so**
- **Turn bad questions into good questions.**

12.4 Using PowerPoint in a Presentation

The usefulness of PowerPoint has come to be questioned in recent years, but there is no denying its continued popularity. It is hard to find a presentation or lecture these days that is not accompanied by PowerPoint slides. A main advantage of PowerPoint over transparency overheads is that the displayed information is visually neater and clearer, giving the impression of professional polish. The main disadvantage of PowerPoint is that unlike those modest transparency overheads and chalkboards of yesteryear, PowerPoint slides can take the audience's eyes and attention away from you and what you are trying to say as presenter, as well as change the attitude of the audience to one of requiring to be entertained and not simply educated. The debate about PowerPoint is ongoing. Used conservatively, however, PowerPoint can be a helpful aid in a vocal presentation. You may be asked to use PowerPoint as part of your presentation. Here are some tips for its use:

- **Choose a sober design template and colour scheme.** You do not want the graphics and artwork of the slides to get in the way of the information you are displaying

- **The design template and colour scheme should be appropriate for the topic you are presenting on.** A harmonious integration of form and content can enhance the effective communication of information and ideas. If you are presenting on cheating in sport, for example, you do not want a design template that includes a fireworks display! Cheating is nothing to celebrate. If you are presenting on relaxation techniques, a design template and colour scheme with cool light blues, aquamarines, and transparent wavelike watermarks would be appropriate, while vibrant yellows, oranges and hot reds would not

- **Avoid animation for entertainment.** The use of animation can be helpful or a hindrance. Used to spice up a dull presentation, it is usually a hindrance. It does nothing to further the pedagogical function of the presentation, and it takes the audience's mind away from the ideas. If you include a lot of animation, you may even become confused as you forget which particular animation trick comes next when you push the key! Don't use animation at all, or use it just to fade or bounce points into the slide. Use the animated underline to highlight important points. Other than these basic, subtle effects, there is really no justification for animation at all in an academic PowerPoint presentation

- **Do not read from the slides.** Use PowerPoint slides to enhance communication. As you give your vocal presentation, bring up the appropriate slides along the way while continuing to maintain eye contact with your audience. The audience can then listen to your details while being able to remind themselves of the main point by glancing at the slide. You should be saying more than is on the slides. The slides are there for the audience, not as your memory aid!

- **Do not use a font smaller than 16 point.** As long as you display only main ideas on your slides, there should be plenty of room using 16 point. It becomes hard for the audience to see what is on the slides if you use a font smaller than this

➢ **Do not pack slides full of information**. A common mistake with PowerPoint is that the slides become the script, packed full of details. PowerPoint slides are a visual aid, and cluttering them with too much information takes away from their function of allowing the audience to keep track of main ideas while focusing on what you have to say about details. Let's take the body checking in minor league ice-hockey topic as our example (see sections 9.1 and 9.2 for more details). Look at the slide below. The slide is packed with information and acts as a script and memory dump for the presenter, but its usefulness to the audience is zero because it contains a dizzying display of visual information.

PERSPECTIVES:

Hockey Canada: Body checking should be allowed in some junior games on a limited, experimental basis, with all safety measures and safeguards in place. Body checking is an important, legitimate skill for controlling the puck, best learned from an early age if safe.
Fans: Body checking is an essential part of the game. It would be impossible to play good ice-hockey without body checking. Ice-hockey would also lose its entertainment value, which would result in a reduced audience, leading to reduced sponsorship, which might endanger the game. Body-checking should be part of ice-hockey.
Scientific Community: The practice of body checking while the brain is still developing is dangerous because damage to the developing brain may not manifest itself until years later.

To correct the slide and let it function for the audience, only the main points should remain. Here is the slide when it has been corrected!

PERSPECTIVES:

❖ **Hockey Canada**

❖ **Fans**

❖ **Scientific Community**

The corrected slides enable members of the audience to listen to the presenter's discussion of facts, figures, and arguments while glancing at the slide to remind themselves of which main point the details are about.